Lecture Notes in Computer Science 16512

Founding Editors

Gerhard Goos
Juris Hartmanis

The series Lecture Notes in Computer Science (LNCS), including its subseries Lecture Notes in Artificial Intelligence (LNAI) and Lecture Notes in Bioinformatics (LNBI), has established itself as a medium for the publication of new developments in computer science and information technology research, teaching, and education.

LNCS enjoys close cooperation with the computer science R & D community, the series counts many renowned academics among its volume editors and paper authors, and collaborates with prestigious societies. Its mission is to serve this international community by providing an invaluable service, mainly focused on the publication of conference and workshop proceedings and postproceedings. LNCS commenced publication in 1973.

Marcin Kowalczyk · Camille Monière
Editors

Design and Architecture for Signal and Image Processing

19th International Workshop, DASIP 2026
Krakow, Poland, January 26–28, 2026
Proceedings

Editors
Marcin Kowalczyk
AGH University of Science and Technology
Krakow, Poland

Camille Monière
Université Bretagne Sud
Lorient, France

ISSN 0302-9743 ISSN 1611-3349 (electronic)
Lecture Notes in Computer Science
ISBN 978-3-032-23870-2 ISBN 978-3-032-23871-9 (eBook)
https://doi.org/10.1007/978-3-032-23871-9

This Springer imprint is published by the registered company Springer Nature Switzerland AG
The registered company address is: Gewerbestrasse 11, 6330 Cham, Switzerland

Preface

This volume contains the papers presented at the 19th Workshop on Design and Architectures for Signal and Image Processing (DASIP 2026), which was held jointly with the 21st HiPEAC Conference in Krakow, Poland, on January 26–28, 2026. The workshop continued its tradition of providing an inspiring international forum for presenting and discussing the latest innovations and developments in the fields of signal, image, and video processing, as well as machine learning, targeting custom embedded, edge, and cloud computing architectures and systems.

DASIP is a long-running annual workshop that was organized for the first time in 2007 in Grenoble, France. Since then, it has been held in numerous locations across Europe and Canada, including Barcelona, Spain (DASIP 2025); Munich, Germany (DASIP 2024); Toulouse, France (DASIP 2023); Budapest, Hungary (DASIP 2022); Montreal, Canada (DASIP 2019); Porto, Portugal (DASIP 2018); Dresden and Munich, Germany (DASIP 2017 and 2024); Rennes, France (DASIP 2016); Krakow, Poland (DASIP 2015); Madrid, Spain (DASIP 2014); Cagliari, Italy (DASIP 2013); Karlsruhe, Germany (DASIP 2012); Tampere, Finland (DASIP 2011); Edinburgh, UK (DASIP 2010); Sophia Antipolis, France (DASIP 2009); and Brussels, Belgium (DASIP 2008).

For this 19th edition of DASIP, 31 papers were submitted from 17 countries around the world. Each paper underwent a rigorous double-blind peer-review process and was evaluated by at least three reviewers selected from the Technical Program Committee. Based on the reviewers' assessments, a selection of 15 high-quality papers was accepted for presentation at the workshop and included in these proceedings.

The success of DASIP 2026 relied on the contributions and dedication of many individuals and organizations. We would like to thank all authors for submitting their work and for contributing to the scientific quality of the workshop. We are also deeply grateful to the members of the Technical Program Committee for their thorough and constructive reviews, which were essential to maintaining the high standards of DASIP. Our thanks also go to the session chairs for their valuable support during the workshop. Finally, we express our sincere appreciation to the members of the Steering Committee for their continued guidance and support.

We also thank Adam Morawiec, from the Institute of Metallurgy and Materials Science, Polish Academy of Sciences, and Paweł Skruch, from AGH University of Krakow, for delivering two insightful keynote presentations that enriched the technical program of DASIP 2026.

January 2026

Marcin Kowalczyk
Camille Monière

Organization

General Chairs

Marcin Kowalczyk	AGH University of Krakow, Poland
Camille Monière	Université Bretagne Sud, France

Steering Committee

João M. P. Cardoso	University of Porto, Portugal
Miguel Chavarrías	Universidad Politécnica de Madrid, Spain
Jean-Pierre David	École Polytechnique de Montréal, Canada
Karol Desnos	INSA Rennes – IETR Laboratory, France
Diana Göhringer	TU Dresden, Germany
Marek Gorgoń	AGH University of Krakow, Poland
Michael Hübner	Brandenburg University of Technology, Cottbus–Senftenberg, Germany
Tomasz Kryjak	AGH University of Krakow, Poland
Paolo Meloni	University of Cagliari, Italy
Sergio Pertuz	TU Dresden, Germany
Sébastien Pillement	University of Nantes – IETR, France
Andrea Pinna	Sorbonne University, France
Alfonso Rodríguez	Universidad Politécnica de Madrid, Spain

Program Committee

Gabriel Caffarena	Universidad CEU San Pablo, Spain
Daniel Chillet	IRISA/ENSSAT University of Rennes, France
Christopher Claus	Robert Bosch GmbH, Germany
Dominique Dallet	IMS/ENSEIRB-Matmeca Bordeaux-INP, France
Martin Danek	Daiteq s.r.o., Czechia
Eduardo de La Torre	Universidad Politécnica de Madrid, Spain
Miloš Drutarovský	Technical University of Košice, Slovak Republic
João Canas Ferreira	University of Porto, Portugal
Guy Gogniat	Lab-STICC/Université Bretagne Sud, France
Bertrand Granado	Sorbonne Université, France
Oscar Gustafsson	Linköping University, Sweden

Frank Hannig	Friedrich-Alexander-Universität Erlangen–Nürnberg, Germany
Kamil Jeziorek	AGH University of Kraków, Poland
Mateusz Komorkiewicz	Aptiv, Poland
Bertrand Le Gal	Inria/ENSSAT University of Rennes, France
Yannick Le Moullec	Tallinn University of Technology, Estonia
Johan Lilius	Åbo Akademi University, Finland
Kevin J. M. Martin	Lab-STICC/Université Bretagne Sud, France
Vincent Migliore	LAAS-CNRS Université de Toulouse, France
Jean François Nezan	IETR/INSA Rennes, France
Arnaldo Oliveira	Universidade de Aveiro/Instituto de Telecomunicações, Portugal
Andrés Otero	Universidad Politécnica de Madrid, Spain
Alessandro Palumbo	Inria Rennes, France
Francesca Palumbo	University of Sassari, Italy
Lazaros Papadopoulos	Democritus University of Thrace, Greece
Maxime Pelcat	IETR/INSA Rennes, France
Christian Pilato	Politecnico di Milano, Italy
Jorge Portilla	Universidad Politécnica de Madrid, Spain
Francesco Ratto	Università degli Studi di Cagliari, Italy
Tiago Rocha	INESC-ID Universidade de Lisboa, Portugal
Nuno Roma	Universidade de Lisboa, Portugal
Olivier Romain	CY Cergy Paris University, France
Claudio Rubattu	University of Sassari, Italy
Ruben Salvador	CentraleSupélec – IETR, France
Hubert Szolc	AGH University of Kraków, Poland
Mateusz Wasala	AGH University of Kraków, Poland

Sponsors

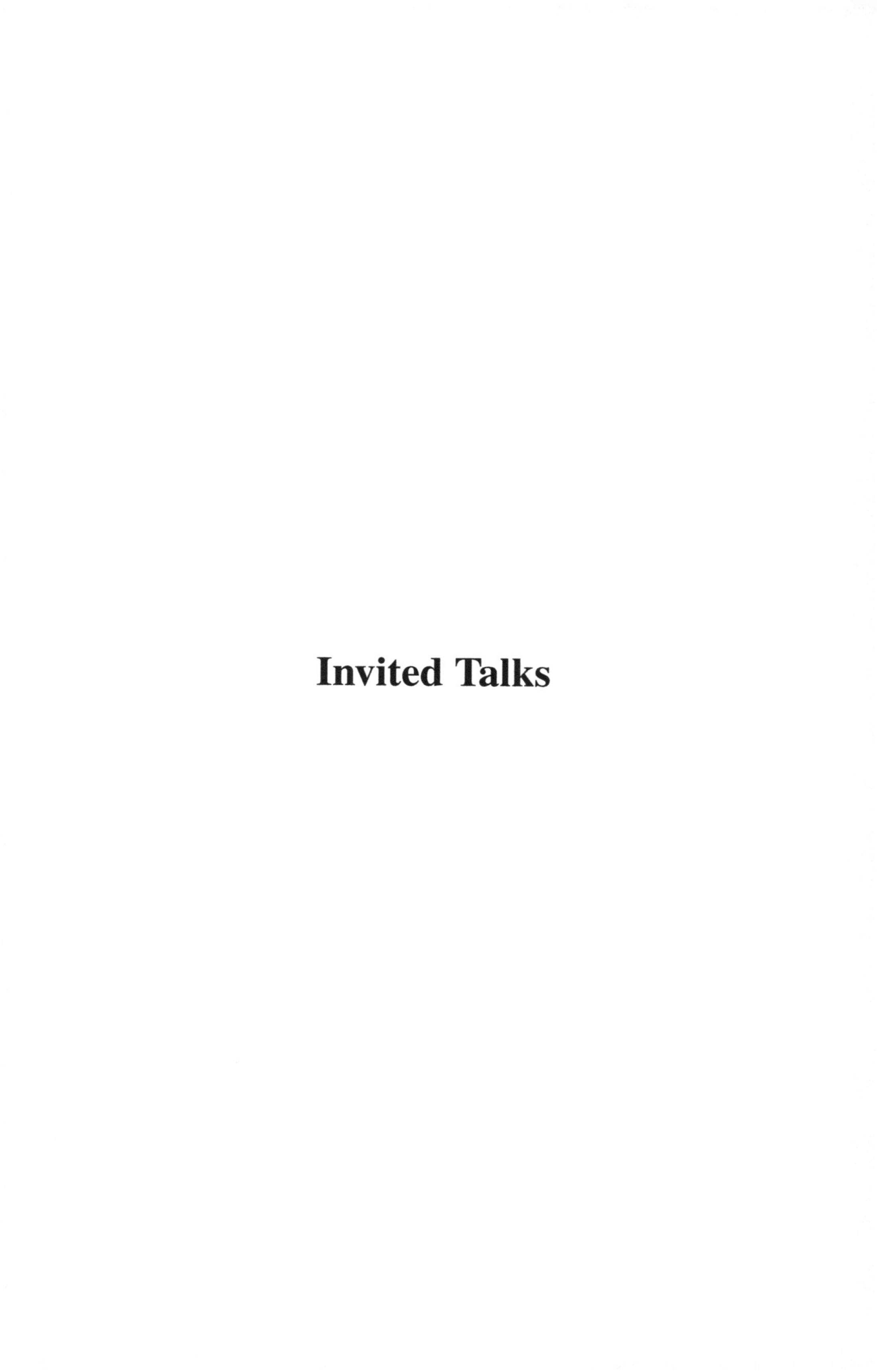

Invited Talks

Network-on-Chip IP for Efficient AI Architectures

Adam Morawiec

Arteris Poland

System-on-Chip (SoC) design complexity has surpassed manual human capabilities, requiring smart interconnect automation based on Network-on-Chips (NoCs). Modern SoCs have 5 to 20+ unique NoC instances and each instance can require 5-10 design iterations in the entire product development. Arteris delivers the data movement IP hardware and IP block integration software to connect in an automated way on-chip components and chiplets. Their use in the design process translates to several advantages: efficient and automated on-chip transport organization enables higher performance in terms of data throughput and latency, lower cost related to interconnect complexity and in-silicon surface and overall lower power consumption. In addition, chip design productivity can be shortened by a factor of up to 10, shortening and reducing iterations from weeks to days. This presentation will provide some insight into these solutions.

The Evolution of Automotive System Architectures in the Era of Software-Defined Vehicles

Pawel Skruch

AGH University of Krakow

The automotive industry is undergoing a profound transformation driven by the transition from hardware-centric designs toward software-defined vehicle (SDV) architectures. This shift fundamentally reshapes how sensing, signal and image processing, decision-making, and control functions are designed, integrated, and deployed within modern vehicles. This keynote traces the evolution of automotive system architectures, from distributed electronic control units (ECUs) with tightly coupled functionalities to centralized and zonal architectures enabled by high-performance computing platforms and software abstraction layers. Particular emphasis is placed on the implications of this evolution for perception systems, where traditional signal- and image-processing pipelines are increasingly replaced or augmented by data-driven and end-to-end learning approaches. The presentation discusses architectural challenges and opportunities arising from SDVs, including sensor fusion, real-time constraints, functional safety, verification and validation, and the coexistence of model-based and learning-based components. Drawing on examples from advanced driver assistance systems and autonomous driving applications, the keynote highlights how architectural decisions influence system performance, scalability, and robustness. The talk concludes by outlining emerging trends and open research directions, positioning software-defined architectures as a key enabler for future intelligent, adaptive, and updatable automotive systems.

Contents

Reconfigurable Computing and Design Tools

A Framework for Automated Integration of High-Level Simulators in RISC-V SoCs

Pedro Ramalho[1,2], Nuno Paulino[1](✉), and João Bispo[1,2]

[1] Faculty of Engineering, University of Porto, Porto, Portugal
{pedro.m.ramalho,nuno.m.paulino,joao.bispo}@inesctec.pt
[2] INESC TEC, Porto, Portugal

Abstract. To achieve further performance and efficiency, SystemonChip (SoC) designs increasingly rely on core customization or integration of application-specific hardware blocks. This requires extensive efforts during Design Space Exploration (DSE) of new hardware to achieve integration, correctness, and target performance. This is time-consuming and error-prone, hindering fast iterative hardware/software co-design. This paper presents a co-simulation framework which integrates arbitrary high-level simulators into Verilog-based SoC platforms, demonstrated on the RISC-V–based open-source X-HEEP SoC. Using inter-process communication we enable cycle-accurate lock-step co-simulation where high-level simulators of accelerators are exposed as memory-mapped peripherals to the RISC-V core. For experimental validation we re-implemented an existing peripheral of the X-HEEP SoC written in Register-Transfer Level (RTL) as an external simulator process, and observed that the co-simulated version maintains identical cycle-level behavior with a maximum wall-clock overhead of 11%. This work enables fast DSEs of heterogeneous RISC-V–based SoCs.

Keywords: RISC-V · Co-Simulation · Hardware acceleration · integration · design-space-exploration

1 Introduction

Modern SoCs increasingly integrate domain-specific accelerators alongside general-purpose Reduced Instruction Set Computer V (RISC-V) cores to meet the performance and energy requirements of emerging workloads. Validating these heterogeneous systems typically relies either on having RTL implementations for all components, or on co-simulation where software models and cycle-accurate RTL components execute together to expose system-level interactions before hardware fabrication. However, existing co-simulation approaches often bind designers to specific modeling languages or require extensive manual effort to connect external high-level simulators with RTL SoC platforms.

These limitations slow down prototyping and hinder fast iteration, especially when exploring alternative accelerator architectures or testing early functional

M. Kowalczyk and C. Monière (Eds.): DASIP 2026, LNCS 16512, pp. 3–15, 2026.
https://doi.org/10.1007/978-3-032-23871-9_1

models. Furthermore, integrating new peripherals frequently demands significant boilerplate around communication interfaces, memory-mapped registers, and synchronization with the simulated SoC. To address these challenges, we propose a framework that automatically integrates high-level accelerator simulators–written in languages such as C++, Python, or Java–into a cycle-accurate RISC-V SoC simulation using Direct Programming Interface (DPI)-based inter-process communication. Our specific contributions are:

1. A declarative JSON-based flow that automatically generates all RTL stubs, DPI interfaces, and simulator templates required to integrate external accelerators.
2. A co-simulation runtime that enables bidirectional communication between Verilated RTL and high-level simulators via ZeroMQ, including support for memory transactions.
3. Validation of the approach via integration with the state-of-the-art X-HEEP SoC, showing cycle-level equivalence and less than 11% wall-clock overhead when replacing hardware accelerators with software models.

The remainder of this paper is organized as follows: Sect. 2 reviews related work in co-simulation and accelerator integration, Sect. 3 presents the architecture and design of the proposed framework, Sect. 4 evaluates its functional accuracy and performance, and finally Sect. 5 concludes the paper and outlines future work.

2 Related Work

We review recent notable simulation and co-simulation works in system-level simulation and hardware/software co-simulation, each targeting the challenges of accurately modeling heterogeneous SoCs and their accelerators.

GVSoC [4] is an open-source, event-driven simulator for PULP-based SoCs that targets high-speed, system-level exploration with accuracy deviations below 10%. The framework models cores, memory, Direct Memory Access (DMA) engines, and peripherals using modular C++ components, while architectural parameters such as interconnect bandwidth and latency are defined through JavaScript Object Notation (JSON) descriptions. Python generators then assemble complete platform instances from these specifications. Component interactions follow a request-based communication model, where memory requests carry address, payload, and latency information and accumulate delays as they propagate through the system. This design enables detailed performance studies without the cost of full cycle-accurate RTL simulation.

ERAS [9] addresses the difficulty of preserving signal-level accuracy when combining RTL models with higher-level, event-driven simulators. Instead of manually bridging these abstraction levels, ERAS uses ESSENT [2] to generate cycle-accurate C-models from RTL netlists, which can be loaded into architectural simulators such as SST or gem5 [3,11]. A parser automatically produces the necessary wrapper code, creating dynamically loadable components that connect the C-models to the simulator's execution flow. This automation significantly reduces integration effort while maintaining accurate low-level behavior, enabling designers to perform architectural exploration without sacrificing RTL fidelity.

FLECSim-SoC [6] is a SystemC-based simulation framework for evaluating accelerators integrated into SoCs, with a focus on Deep Neural Network (DNN) co-design. The approach models Central Processing Units (CPUs), memories, and accelerators as SystemC components connected through a Transaction-Level Modeling (TLM)-based crossbar, and supports the inclusion of cycle-accurate RTL modules via Verilator. System architecture is defined in JSON files, enabling designers to vary parameters such as memory hierarchy, Dynamic Random Access Memory (DRAM) timing, and DMA behavior. The framework provides both instruction-level processor models and detailed accelerator interactions, and it integrates with energy and area estimation tools such as Accelergy [13], making it suitable for exploring hardware–software trade-offs in DNN workloads.

RoSÉ [10] is a hardware/software co-simulation infrastructure that enables closed-loop evaluation of robotics systems by coupling an environment simulator called AirSim [12] with the cycle-accurate RTL simulation framework FireSim [7]. AirSim generates sensor data and processes actuation commands at discrete time steps, while FireSim executes the target SoC at clock-cycle granularity on Field-Programmable Gate Arrays (FPGAs). A dedicated bridge module synchronizes the two domains by exposing memory-mapped Input/Output (I/O) registers and exchanging state information over Transmission Control Protocol (TCP). A lockstep synchronizer aligns simulation time across both engines–despite their differing time bases–ensuring that sensor updates, actuator commands, and hardware events remain coherent throughout execution. This enables realistic pre-silicon evaluations of robotics workloads with tight hardware–environment interaction fidelity.

Kwon *et al.* [8] propose a simulation framework that couples QEMU-based software emulation with real-time RTL models through a virtual interface layer. QEMU generates a JSON-described register map and is extended to forward peripheral register activity to Verilog simulations via IEEE PLI [1], while a socket-based communication module enables bidirectional data exchange and debugging over TCP. The framework supports dynamic loading of RTL modules and distributes simulation workloads across multiple hosts by segmenting models into parameterized units. This design allows embedded software to interact seamlessly with cycle-accurate hardware blocks, enabling flexible architectural exploration without modifying application code.

Busnot *et al.* [5] introduce a parallel SystemC kernel that accelerates loosely-timed TLM simulations while remaining compliant with the IEEE SystemC standard. The kernel evaluates SystemC processes concurrently on multicore hosts through a parallel scheduling mechanism that preserves atomicity and determinism, avoiding race conditions. It also supports Direct Memory Interface (DMI), allowing simulated components to access host memory directly and reducing communication overhead typical of higher-level messaging. Although loosely-timed TLM abstractions sacrifice fine-grained synchronization, they enable fast system-level exploration, and the addition of DMI further improves performance when coupling hardware models with external software processes.

Although these frameworks improve system-level and co-simulation capabilities, they are typically bound to specific modeling environments and require manual effort to integrate new accelerator models. None provide a generic, language-agnostic way to connect external high-level simulators to a cycle-accurate RISC-V SoC. Our approach instead automates this integration through a declarative specification and DPI-based inter-process communication, enabling rapid prototyping with minimal boilerplate.

3 Co-simulation Framework Architecture

The framework is structured around three components that work together to integrate high-level simulators into a cycle-accurate RISC-V SoC. First, the hardware platform provides the simulated SoC environment, where peripherals appear as standard memory-mapped modules connected to the processor. Second, each peripheral has an associated high-level simulator process, written in a High-Level Language (HLL), which implements all functional behavior while exchanging register values and memory requests with the SoC. Third, a code-generation flow automates integration by producing the RTL stubs, DPI interfaces, and simulator templates from a concise JSON specification.

Figure 1 shows a concrete instantiation of the proposed simulation architecture using X-HEEP, an open-source SoC implemented in SystemVerilog (SV) that integrates a RISC-V processor core, memory system (omitted), and peripheral subsystem which includes already commonly used peripheral modules such as I2C or UART (omitted). A host application is compiled and executed on the RISC-V within X-HEEP, which as a whole is simulated via Verilator. The host code interacts with peripherals using Memory-Mapped Input/Output (MMIO), among which are accelerator modules whose connection to the bus is implemented via SV wrappers which contain C-based DPI calls acting as the interface to a ZeroMQ based message passing step which connects to external processes. In summary, a RISC-V binary is executed on the simulated X-HEEP processor, while external processes implement the accelerator behavior (exemplified in Listing 1.3).

3.1 Hardware Platform Integration

The hardware platform provides the cycle-accurate SoC environment in which all co-simulated peripherals operate. In this setting, each accelerator appears to the RISC-V core as a standard memory-mapped SV module connected through MMIO. Each generated peripheral exposes two interfaces to the SoC: an MMIO register file for control and status, and a pair of Open Bus Interface (OBI) channels for issuing memory read and write transactions. Although all functional

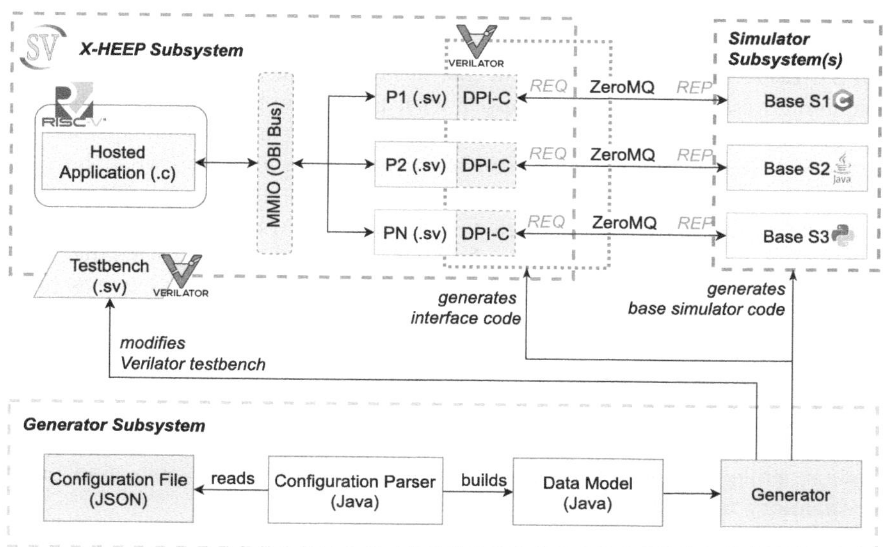

Fig. 1. Co-simulation framework instantiated on the X-HEEP RISC-V SoC. Boilerplate code to support the co-simulation is generated from JSON configuration files. The host application executes on the simulated RISC-V core, while accelerator functionality is implemented in external high-level simulator processes.

behavior is delegated to an external simulator process, the peripheral remains an ordinary RTL block. It includes clock and reset signals, address-decode logic, and an OBI-compatible master interface so the simulator can initiate memory operations through the SoC interconnect.

To bridge the SV environment with the high-level simulator, each generated peripheral's top-level module uses DPI-C calls to route register values and memory-access results between the RTL domain and a C layer linked into the Verilated SoC simulator. Listing 1.1 shows a minimal example of an MMIO-mapped peripheral whose registers are exchanged with an external simulator.

Listing 1.1. Exerpt of the communication flow in generated SV peripheral wrappers.

```
// (code omitted)
import "DPI-C" function void communication_send(int r1, int r2);
import "DPI-C" function void communication_recv(int r3);
// DPI declarations will link with C-side simulation

always_ff @(posedge clk_i or negedge rst_ni) begin
    if (rst_ni)
        int signed _r3; /* temporary variable... */
        communication_send(r1, r2, ...); /* send inputs... */
        communication_recv(_r3); /* recv outputs... */
        r3 <= _r3;
    end
end
```

3.2 Simulator Processes

Each hardware module or accelerator instantiated in the SoC as described above, implemented as a simple RTL wrapper with DPI calls, has an associated simulator process written in C++, Python, or Java. The simulator encapsulates all functional behavior of the accelerator and executes in lock-step with the cycle-accurate RTL model of the SoC. At every cycle, the simulated SoC sends the peripheral's MMIO-visible registers to the simulator process, and the simulator returns updated register values. As mentioned above, these exchanges pass through an intermediate abstract controller interface, which also exposes blocking read and write operations implemented in the generated base-class templates, and these correspond directly to the peripheral's OBI master channels. Thus, the simulator may trigger memory transactions without manipulating low-level protocol signals. This design makes the simulator responsible for the entire datapath logic, including state updates, FIFO management, scheduling, and algorithmic behavior, while the hardware wrapper only forwards register values and propagates memory requests to the SoC interconnect.

One simulator process is spawned per peripheral instance, and these processes are independent. This is important for scalability, because parallel simulators can model heterogeneous accelerators or multiple copies of the same accelerator without modifying the SoC testbench. Because the simulators run in user space and communicate through ZeroMQ over DPI-C, they can be modified and recompiled independently of the Verilated SoC, which significantly reduces iteration time. Since all communication between the RTL wrapper and the simulator occurs through this DPI-C based channel which is integrated in the Verilator testbench, this facilitates implementation of future features for tracing, assertions, and debugging.

3.3 Code-Generation Flow

The code-generation flow automates all steps required to integrate a new peripheral and its simulator into the SoC environment. The entry point is a JSON specification that declares the peripheral layout. This includes the memory base address, register file, data types, directions, and any configuration parameters required by the simulator.

This JSON file is parsed into an internal data model and processed by an emitter pipeline that generates three categories of artifacts: 1) a SystemVerilog peripheral wrapper that exposes MMIO registers and DPI-C entry points, 2) a simulator project for C++, Python, or Java that includes base classes and a controller interface that abstracts OBI reads and writes, and 3) integration files for the SoC testbench such as memory map definitions and build rules. Besides generating these files, certain build files need to be updated, which the current generator also supports, when the target is the X-HEEP platform.

Listing 1.2 shows a simplified example of the JSON block used to generate the module we employ for experimental evaluation in Sect. 4. The snippet shows how the user specifies only the peripheral identifier, target language, memory region and register definitions.

Listing 1.2. Peripheral definitions block of replicated `simple_accelerator` example.

```
{ "identifier": "replicated_simple_accelerator",
  "description": "A simple accelerator",
  "class_name": "replicated_simple_accelerator",
  "target_language": "cpp",
  "hardware_interface": {
    "memory_base_address": 6000,
    "memory_size_bytes": 100,
    "registers": [
      {
        "name": "src_address",
        "direction_to_simulator": "input",
        "type": "logic",
        "signed": false,
        "width": 32,
        "default_value": 0
      },
      // <... omitted other registers and OBI bus interface ...>
    ]
  }}
```

Based on this specification, the full SystemVerilog wrapper, DPI interface and simulator base class which inherits communication interfaces for register and OBI bus access. The developer then fills in only the domain logic, as illustrated in the shortened example of Listing 1.3.

Listing 1.3. A C++ implementation of a simple accelerator's domain-specific logic.

```
class simple_accelerator : public base_simple_accelerator {
  // (code omitted)
  std::queue<write_req> fifo;
  simulator_state busy(const sim_data &in, sim_data &out) override {
  /* busy method called from FSM in parent class*/

    if (!fifo.empty()) { /* process write requests first */
      auto &[addr, data] = fifo.front();
      if (this->controller.single_write(addr, data, in, out)) {
        fifo.pop();
        wr_count++;
      }}

    if (rd_count < this->get_data_size()) { /* start issuing reads */
      uint32_t src_addr = this->get_src_address() + rd_count * WORD_SIZE;
      uint32_t dst_addr = this->get_dst_address() + wr_count * WORD_SIZE;
      int32_t v_actual = this->controller.single_read(src_addr, in, out);
      if (v_opt.has_value()) {
        int32_t v_filtered = v_actual > this->get_threshold()
            ? v_actual : this->get_threshold();
        fifo.push({dst_addr, v_filtered});
        rd_count++;
      }}

    if (fifo.empty() && /* everything has been read and written */
      rd_count == this->get_data_size() && wr_count == this->get_data_size())
      return simulator_state::DONE;
```

```
        return simulator_state::BUSY;
    }
};
```

The above example follows a request–response execution flow in which the simulator's `busy()` method is invoked once per cycle. The base class handles serialization, register transfer and memory-transaction state machines, ensuring that simulator and RTL wrapper remain synchronized. Users only implement the algorithm-specific logic in the simulator.

Table 1. Experimental setup specifications

Component	Specification
CPU	Intel(R) Core(TM) i5-9400F CPU @ 2.90GHz
RAM	2x8GiB DIMM DDR4 Synchronous Unbuffered 2667 MHz
GCC/G++	11.4.0
RISC-V Toolchain	Cloned from branch "2022.01.17"
Verilator	4.210 2021-07-07 rev v4.210

4 Experimental Evaluation

We evaluate the proposed framework by replicating and extending an existing accelerator example from the X-HEEP SoC. This accelerator applies a threshold filter to a buffer of data and writes the filtered results to another memory location using OBI transactions. The original Intellectual Property (IP) example is written entirely in SV and resides in the X-HEEP peripheral subsystem as a memory-mapped component, with a host application controlling its execution.

We performed two experiments: (i) to verify that a co-simulated accelerator produced by our code-generation flow can reproduce the functional and cycle-level behavior of the original SystemVerilog implementation, and (ii) to measure the performance overhead introduced by the communication layer when scaling the number of accelerator instances.

The X-HEEP microcontroller was generated with default settings except for increasing the number of memory banks to 12 to support the large buffers used in our tests. All experiments run within a Verilated X-HEEP simulation using identical host software, enabling a direct comparison between the original accelerator and its co-simulated counterpart. Other hardware and software components of our experimental runs are presented in Table 1.

4.1 Experiment 1: Replicating the Simple Accelerator

Replicating the `simple_accelerator` example with our framework is a straightforward process that minimizes boilerplate and manual integration effort. The first step is to write a JSON configuration file defining the four registers used by the threshold-filter algorithm (source and destination addresses, threshold, and data size). Once the configuration file is processed, the generator produced all necessary artifacts for X-HEEP and the simulator. Namely, the SV peripheral stub for integration into the SoC and base simulator project. C++ is chosen as the target simulator language for this experiment. The behavior of the simulated peripheral is controlled by the host program on the RISC-V core present in the SoC, which initializes the accelerator, waits for completion, and then verifies correctness by comparing every element of the output buffer against the expected filtered values, returning the number of mismatches as the program's exit code.

Results. Figure 2 presents the number of clock cycles required to complete the simulation for various data sizes, comparing the original and replicated implementations. For all data sizes except the smallest one ($N = 2$), both implementations perform identically in terms of cycle count. At $N = 2$, the replicated version requires an additional 100 cycles (2102 vs. 2002). This difference in cycles probably indicates that the C++ implementation of the accelerator might be missing some edge-case. However, for every other value of N, simulation times were identical.

Figure 2 plots the ratio between the wall-clock times of the replicated and the original implementations. We observe a rather consistent overhead, between 6% (for $N = 2$) and 11% (for $N = 64$). Considering that the data sizes are increasing exponentially, the overhead introduced by the simulation framework appears to be constant in relation to the total simulation time.

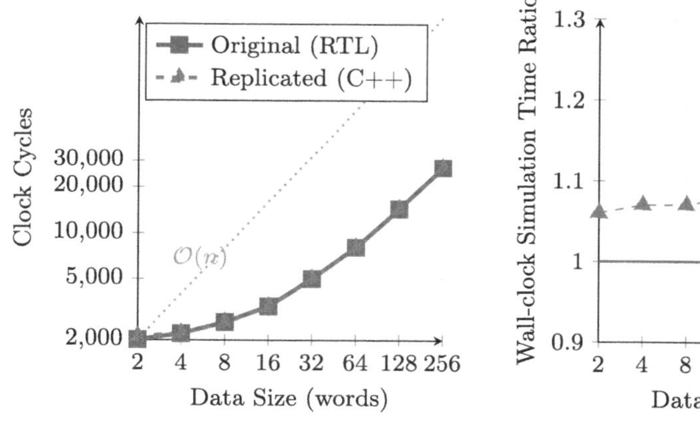

(a) Simulation framework cycle accuracy (b) Overhead of simulation framework

Fig. 2. Single accelerator results for different data sizes implementations.

4.2 Experiment 2: Scaling for Multiple Accelerators

The second experiment evaluates how well our framework handles simulation workloads when scaling the number of instantiated accelerators within a single system. To do this, we replicate the same accelerator used in the previous experiment and instantiate multiple copies in the same simulation environment. This approach is used for both the original accelerator, written purely in SV, and the replicated `C++` version. We report results for configurations with 2, 4, and 8 accelerators running concurrently, all using a fixed buffer size of 256.

Results. Figure 3 plots the growth of simulation cycles for buffer sizes of 256, using 2, 4, and 8 accelerator instances. Both implementations achieved the exact same number of simulation clock cycles. Figure 3 presents the ratio between the wall-clock simulation times of the two versions. The overhead remains under 10%, regardless of the number of accelerator instances. These results are generally consistent with those from the first experiment, indicating that the framework induces a constant overhead that does not grow disproportionately with the number of accelerator instances.

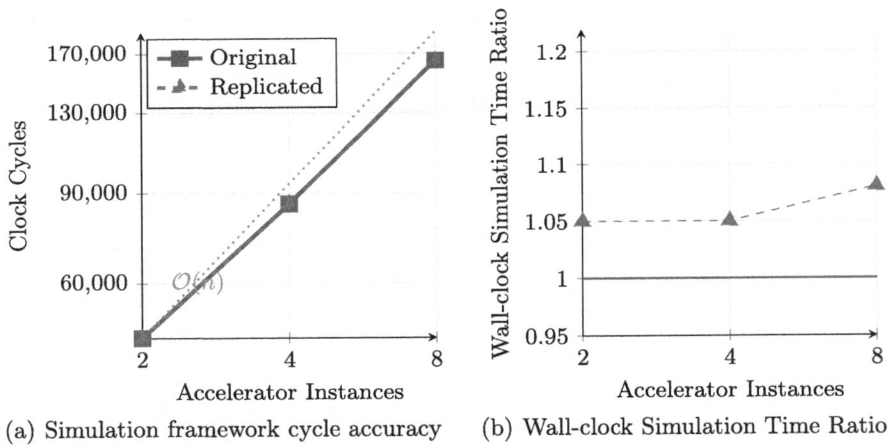

(a) Simulation framework cycle accuracy (b) Wall-clock Simulation Time Ratio

Fig. 3. Single accelerator results for different number of accelerators.

4.3 Analysis

In terms of simulation cycles, the only observed deviation occurred at $N = 2$, during the first experiment, where a difference of 100 cycles was recorded. While this technically disqualifies the implementation from being cycle-accurate across all cases, we believe this discrepancy stems from minor timing differences that the `C++` version did not manage to accurately capture. Given that all other tested configurations produced identical cycle counts, we are confident that the implementation remains a very close approximation of the intended behavior.

Wall-clock times for the replicated versions were consistently higher, but the overhead remained modest, reaching a maximum of around 11%. More importantly, this overhead is stable and does not increase disproportionately with the workload. Despite the exponential increase in data value, the overhead curve remains flat, indicating that the replication framework imposes a roughly constant cost rather than a scaling penalty.

Another important advantage, not reflected in these results, is the reduction in turnaround time when modifying the simulator. Unlike traditional flows that require costly regeneration and recompilation of the entire SoC simulator (for example, with Verilator), in our framework, changes to the replicated peripheral model can be tested rapidly without recompilation of the whole system. This aligns well with the framework's goal of providing an agile prototyping and testing environment for peripherals integrated in complex systems.

It is worth noting that the accelerator used in these experiments is intentionally simple, with minimal input requirements and trivial domain logic. While this setup serves as a reasonable initial test, it may represent a worst-case scenario for co-simulation approaches where the workload is dominated by data transfer rather than computation. In such cases, higher-level implementations (for instance, in `C++` or Java) may even outperform cycle-accurate RTL simulation. Therefore, further testing with more complex and computation-heavy accelerators is necessary to fully evaluate scalability and robustness.

Nonetheless, we consider that these proof-of-concept results validate the approach. Overall, the framework's architectural overhead is manageable and consistent with the trade-offs made during design, while offering practical benefits in simulation workflow efficiency.

5 Conclusion

This work presented a co-simulation framework that integrates high-level accelerator models into cycle-accurate RISC-V SoCs through a declarative specification and automated code-generation flow. By combining generated RTL stubs, DPI-based communication, and message-oriented interaction via ZeroMQ, the approach enables high-level simulators to function as memory-mapped peripherals and autonomously issue memory transactions within a Verilated SoC environment. The experimental evaluation using the X-HEEP platform demonstrated cycle-accurate behavior, incurring at most an 11% wall-clock time overhead. These results indicate that the framework provides a practical and efficient path for rapid prototyping and iterative refinement of accelerator designs without requiring full RTL re-implementation or costly rebuilds of the entire simulation environment.

Future work will extend the experimental evaluation to more complex accelerator architectures, such as spatial arrays, as well as tightly coupled integration cases, as we have recently implemented support for co-simulation of custom instruction extensions through the Core-V eXtension interface (CV-X-IF) bus, allowing us to study instruction-level acceleration and realistic application workloads beyond memory-mapped peripherals.

Acknowledgments. This work has received funding within the Chips Joint Undertaking (Chips JU) - the Public-Private Partnership for research, development and innovation under Horizon Europe – and National Authorities under GA 101096658, and by National Funds through the FCT - Fundação para a Ciência e a Tecnologia, I.P. (Portuguese Foundation for Science and Technology) within the project AIMaCoV, with reference 2023.14899.PEX (https://doi.org/10.54499/2023.14899.PEX).

References

1. IEEE Standard for Verilog Hardware Description Language. IEEE Std 1364-2005 (Revision of IEEE Std 1364-2001), pp. 1–590 (2006). https://doi.org/10.1109/IEEESTD.2006.99495
2. Beamer, S., Donofrio, D.: Efficiently exploiting low activity factors to accelerate RTL simulation. In: 2020 57th ACM/IEEE Design Automation Conference (DAC), pp. 1–6 (2020). https://doi.org/10.1109/DAC18072.2020.9218632
3. Binkert, N., et al.: The gem5 simulator. SIGARCH Comput. Archit. News **39**(2), 1–7 (2011). https://doi.org/10.1145/2024716.2024718
4. Bruschi, N., Haugou, G., Tagliavini, G., Conti, F., Benini, L., Rossi, D.: GVSoC: a highly configurable, fast and accurate full-platform simulator for RISC-V based IoT processors. In: 2021 IEEE 39th International Conference on Computer Design (ICCD), pp. 409–416. https://doi.org/10.1109/ICCD53106.2021.00071
5. Busnot, G., Sassolas, T., Ventroux, N., Moy, M.: Standard-compliant parallel SystemC simulation of loosely-timed transaction level models. In: Proceedings of the Asia and South Pacific Design Automation Conference, ASP-DAC, vol. 2020-January, pp. 363–368 (2020). Institute of Electrical and Electronics Engineers Inc. https://doi.org/10.1109/ASP-DAC47756.2020.9045568
6. Hotfilter, T., Hoefer, J., Kreß, F., Kempf, F., Becker, J.: FLECSim-SoC: a flexible end-to-end co-design simulation framework for system on chips. In: 2021 IEEE 34th International System-on-Chip Conference (SOCC), pp. 83–88 (2021). https://doi.org/10.1109/SOCC52499.2021.9739212
7. Karandikar, S., et al.: FireSim: FPGA-accelerated cycle-exact scale-out system simulation in the public cloud. In: 2018 ACM/IEEE 45th Annual International Symposium on Computer Architecture (ISCA), pp. 29–42 (2018). https://doi.org/10.1109/ISCA.2018.00014
8. Kwon, J., Oh, S., Park, D.: Metamorphic edge processor simulation framework using flexible runtime partial replacement of software-embedded verilog RTL models. In: Proceedings - IEEE International Symposium on Circuits and Systems (2021). Institute of Electrical and Electronics Engineers Inc. https://doi.org/10.1109/ISCAS51556.2021.9401354

9. Nema, S., et al.: ERAS: a flexible and scalable framework for seamless integration of RTL models with structural simulation toolkit. In: Proceedings - 2023 IEEE International Symposium on Workload Characterization, IISWC 2023, pp. 196–200. Institute of Electrical and Electronics Engineers Inc. https://doi.org/10.1109/IISWC59245.2023.00038
10. Nikiforov, D., Dong, S.C., Zhang, C.L., Kim, S., Nikolic, B., Shao, Y.S.: RoSÉ: A Hardware-software co-simulation infrastructure enabling pre-silicon full-stack robotics soc evaluation (2023). https://doi.org/10.1145/3579371.3589099
11. Rodrigues, A.F., et al.: The structural simulation toolkit. **38**(4), 37–42 (2011). https://doi.org/10.1145/1964218.1964225
12. Shah, S., Dey, D., Lovett, C., Kapoor, A.: AirSim: high-fidelity visual and physical simulation for autonomous vehicles. In: Field and Service robotics: Results of the 11th International Conference, pp. 621–635. Springer, Cham (2017). https://doi.org/10.1007/978-3-319-67361-5_40
13. Wu, Y.N., Emer, J.S., Sze, V.: Accelergy: an architecture-level energy estimation methodology for accelerator designs. In: 2019 IEEE/ACM International Conference on Computer-Aided Design (ICCAD), pp. 1–8 (2019). https://doi.org/10.1109/ICCAD45719.2019.8942149

An Open-Source Framework Bridging System-Level Measurements and Circuit-Level Energy Consumption and Latency Characteristics

Johannes Knödtel(✉) and Marc Reichenbach

Institute of Applied Microelectronics and Computer Engineering, University of Rostock, Rostock, Germany
{Johannes.Knoedtel,Marc.Reichenbach}@uni-rostock.de

Abstract. Evaluating non-functional properties such as power consumption on microcontroller platforms exposes a gap between abstraction layers: while programs execute at the Instruction Set Architecture (ISA) level, energy consumption is determined at the circuit level. This connection between system and circuit layers is especially relevant when modeling non-functional processor properties. Modern approaches (e.g., deep neural networks) rely on large datasets, which in turn require automated methods that generate consistent labelling between programs and their measured energy consumption and latency. We introduce an open-source, automated framework that applies a measurement-based approach for energy consumption and latency characterization. The framework allows users to treat hardware as a physical black box while still obtaining measurements of instruction sequences, even without performance counters or integrated power-monitoring infrastructure. The system is built entirely from low-cost commercial off-the-shelf (COTS) components. It relies on repeated measurement cycles, averaging many iterations of the same program under test. Each cycle includes a state-flush routine, which executes code intending to reset the microarchitectural state to capture state-dependent effects. Preliminary experiments on two RISC-V platforms (ESP32-C6 and BananaPi BPi-F3) reveal that some of the tested state-flush methods introduce a high degree of unwanted noise, especially for energy measurements.

1 Introduction

Energy efficiency and related non-functional properties are first-class design constraints in embedded systems. Designing for these criteria is increasingly critical, yet remains highly challenging. A central difficulty lies in the lack of visibility into circuit-level behaviour. System designers rarely operate at this level and, in most cases, cannot: information beyond the ISA, such as Register Transfer Level (RTL) sources or layouts, is often proprietary. Even with open-source hardware, such as the PULP line of processors [2], meaningful insights require

M. Kowalczyk and C. Monière (Eds.): DASIP 2026, LNCS 16512, pp. 16–28, 2026.
https://doi.org/10.1007/978-3-032-23871-9_2

a full Integrated Circuit (IC) flow, proprietary Process Design Kits (PDKs), and simulations that are often prohibitively time-consuming or infeasible. Consequently, researchers and designers must treat processors as black boxes and rely on experimental methods.

Beyond raw measurements, meaningful analysis requires datasets that systematically label programs with energy consumption and latency, forming the basis for modeling and optimization. Such datasets are valuable both for predictive models and for compiler optimizations. This directly implies the need for a framework that can automatically and reproducibly execute programs and assign energy consumption and latency values. While this need is well recognized in embedded systems research, existing testbeds typically focus on measurement alone. What is missing is a framework that unifies the following requirements:

1. **End-to-end automation**: capable of measuring arbitrary programs without human intervention,
2. **Full openness**: fully open-source and reproducible by other research groups,
3. **Low cost and accessibility**: relying solely on COTS components.

In this paper we present a first proof of concept for such a framework, providing the community with a practical, reproducible, and extensible solution. An overview of the concept is shown in Fig. 1.

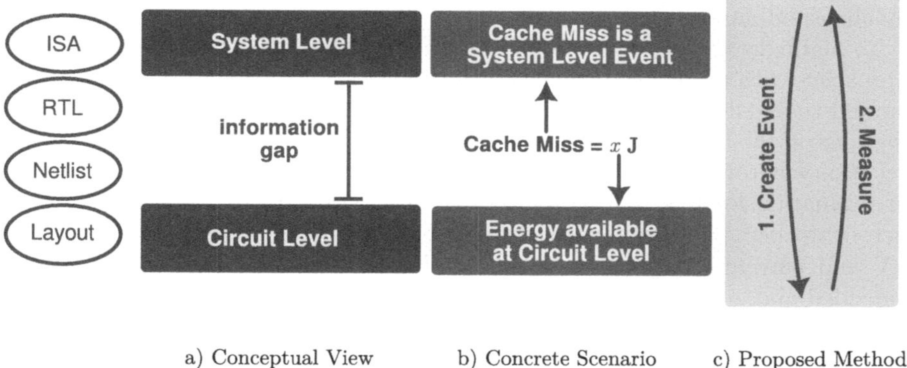

Fig. 1. Graphical summary of this work.

The contributions of this work are threefold:

- We present an open-source, low-cost framework that automates end-to-end measurement cycles, including program upload, configurable state-flush routines, controlled execution, and collection of current/voltage traces.
- We provide a portable test harness with backends for both a bare-metal microcontroller platform and generic Linux-based systems, and demonstrate it on two RISC-V platforms: an ESP32-C6 microcontroller and a BananaPi BPi-F3 single-board computer.

- We conduct a first systematic evaluation of state-flush strategies and their impact on energy and latency measurements, highlighting non-trivial noise and bias effects that call for careful design of microarchitectural reset procedures.

The remainder of this paper is structured as follows: Sect. 2 reviews related approaches. Section 3 describes the hardware and software components of our system as well as extensibility to other platforms. Section 4 presents an experimental evaluation of the framework and the state-flush mechanism. Section 5 concludes the paper.

2 Related Work

This section reviews notable similar approaches and positions our framework in relation to them. Most prior efforts concentrate on the measurement technique itself, whether targeting energy, latency or both, whereas our work emphasizes the broader perspective of a generic, automated measurement loop.

In particular, existing methods rarely integrate the steps required for systematic dataset creation. Tasks such as clearing microarchitectural states, uploading programs, and iterating over a benchmark corpus are often left to manual effort or ad-hoc scripts. Our framework incorporates these elements end-to-end, enabling not only measurements but also reproducible dataset generation for modeling and analysis.

To highlight the position of our framework, Table 1 contrasts representative approaches with respect to openness, cost, and degree of end-to-end automation. Our end-to-end flow is extensible by implementing a small set of platform-specific functions on the DUT. This abstraction allows the framework to adapt to diverse execution environments while preserving a high degree of automation. Although our evaluation focuses on two RISC-V devices, the approach is not limited to microcontrollers. The supported current sensors tolerate supply voltages up to 36 V and currents up to 15 A, enabling the measurement of a wide range of embedded and general-purpose platforms.

The main practical requirement is a start/stop trigger signal that marks the measurement window. While this can be implemented via a GPIO on many systems, some platforms, such as compact x86_64 single-board computers, offer limited GPIO accessibility. In such cases, hardware-level triggering remains possible by probing alternative signals (e.g., activity LEDs, debug headers, or firmware-exposed pins) or by using on-chip debug interfaces that can emit external trigger pulses.

The framework can also be applied to soft-core processors on FPGAs. Here, the DUT can generate dedicated trigger signals directly in hardware, enabling precise and deterministic measurement windows. This makes the methodology suitable for evaluating different microarchitectures instantiated on reconfigurable logic as well as a broad class of embedded and general-purpose devices.

Nonetheless, our primary focus is on embedded devices, where energy consumption and latency are critical design constraints. The trade-offs in our work, compared to prior approaches, lie primarily in the measurement specifications. Because we operate the current sensor close to the lower end of its measurable range, the quantization step becomes significant, resulting in noticeable quantization noise. Furthermore, our effective temporal resolution is lower than that of high-end instruments; based on empirical observations, we estimate a time resolution of approximately 0.25 kHz. On the other hand, the total cost of the components used in our setup was below 50 €, which makes the framework inexpensive to reproduce. This low cost enables multiple identical setups

Table 1. Measurement-based approaches and end-to-end capability (OS = open source, ✓= Yes, • = Partial, ×= No; Cost (estimated): L: <100€, M: 100€–500€, H: >500€).

Name	Target scenario	OS	Cost	Automation
EMPIOT (2018) [3]	Energy measurement of IoT and wireless devices	•	M	**Partial**: measure+log; no upload
[11]	Low-cost energy measurement of various devices	×	M	**Weak**: focus on measurement
SES (2002) [1,10]	Energy analysis platform for ARM7TDMI	×	M	**Very high**: tightly integrated analysis platform
FlockLab (2013) [6]	Testbed for wireless embedded systems	✓	H	**Strong**: very high integration
Energino (2012) [4]	Low-cost Arduino power logger for wireless applications	✓	L	**Weak**: measurement only
CurrentRanger [9]	Precision ammeter	✓	M	**Weak**: only logging
Joulescope [5]	USB profiler (nA–A, GUI/API)	•	M	**Partial**: APIs, but no integration with Device under Test (DUT)
Nordic PPK2 [7]	Power Profiler with focus on low power	×	L	**Strong**: vendor-specific application, integration into development kits
PowerSensor3 (2025) [12]	Power profiler with focus on HPC (GPUs, FPGAs, etc.)	✓	M	**Strong**: powerful APIs
Otii Arc Pro / Ace Pro [8]	High-resolution power profilers (0.5–25 V, nA–5 A range, up to 50 ksps)	×	H	**Partial–Strong**: synchronized measurement; no upload/state flush
This work (2025)	Automated energy/latency characterization	✓	L	**Strong**: end-to-end upload, state flush, and measurement loop

to be deployed in parallel, facilitating large-scale or distributed experimental campaigns.

3 Methodology

We now describe the hardware and software components of the proposed framework. The hardware setup consists of a controller for acquisition and automation (Raspberry Pi Zero 2 W), a current/voltage sensor (currently supported: Texas Instruments INA219/226/260), and the device under test (DUT), which in our evaluation is either an ESP32-C6 microcontroller or a BananaPi BPi-F3 single-board computer. An overview is shown in Fig. 2.

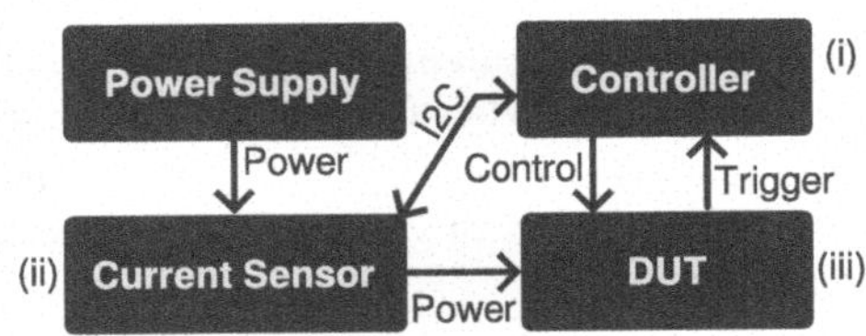

Fig. 2. Hardware setup of the measurement framework.

At the core of the system is an automated measurement loop designed to obtain statistically reliable, repeatable measurements. Each cycle begins with transferring the program under test to the DUT. Once the DUT signals readiness via a General-Purpose Input/Output (GPIO) line, the controller begins recording current and voltage. Before the actual test program executes, the DUT performs a state-flush routine intended to reset microarchitectural state. After a fixed user-defined execution window, the DUT signals completion and the controller terminates the measurement. The overall workflow is illustrated in Fig. 3.

A central component of this process is the *state-flush routine*, which combines several mechanisms targeting different microarchitectural structures. We refer to these mechanisms as `CACHE`, `BP`, `BTB`, `ICACHE_SYNC`, and `PIPE_DRAIN`. `CACHE` applies irregular-stride memory accesses to disturb data caches and prefetchers. `BP` introduces unpredictable conditional branches to perturb the branch predictor, while `BTB` disrupts branch-target-buffer entries. `ICACHE_SYNC` issues an explicit `fence_i` to enforce instruction-cache consistency, and `PIPE_DRAIN` executes sequences of `nop` instructions to drain the pipeline. Their combined purpose is to reduce the influence of prior executions on subsequent measurements. The trade-off between the noise they introduce and the consistency they provide is evaluated in Sect. 4.

To isolate the overhead of the flush routine, the framework performs an additional "empty" measurement in which only the state-flush is executed and the per-iteration values are subtracted from the actual run. The measured time and energy of this run are subtracted from those of the test run.

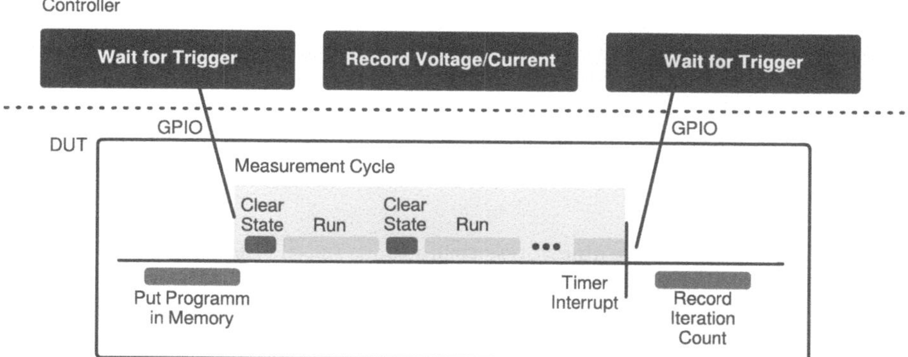

Fig. 3. Automated measurement cycle.

Beyond the test harness used in this paper, the framework was designed with portability in mind. Support is currently implemented for Linux-based devices and for the ESP32 family of microcontrollers. Adding additional platforms requires implementing a small set of platform-specific functions that define how programs are transferred, executed, and how the microarchitectural reset is triggered. In the evaluation, we instantiate this interface for a Linux-based RISC-V single-board computer (BananaPi BPi-F3) and for an ESP32-C6 microcontroller. These required functions are summarized in Table 2; they encapsulate the device-specific aspects of execution so that the remainder of the framework remains platform-agnostic.

Table 2. Functions to be implemented for supporting a new platform

Name	Functionality
`accept_client`	Accept network connection
`close_client`	Close network connection
`alloc_exec`	Allocate executable memory
`free_exec`	Free executable memory
`run_program_loop`	Execute program under test
`microarch_reset_kernel_begin`	Execute state-flush routine
`on_cycle_begin`	*Optional:* Pre-measurement set-up
`on_cycle_end`	*Optional:* Post-measurement clean-up

Although the focus of this work is on energy and latency, the framework can also be used to measure throughput. In practice, the per-iteration latency already provides an inverse measure of throughput, but a more direct evaluation is possible by disabling the state-flush routine and allowing the program under test to execute continuously in a tight loop. This configuration removes the per-cycle reset overhead and therefore reflects the sustained throughput of the workload itself. We do not analyze throughput in detail in this paper, but the framework could be utilized in this way simply through configuration.

4 Evaluation

This section evaluates the framework, with a focus on the behaviour and impact of the state-flush mechanism. We begin by describing the hardware platforms, test cases, and evaluated configurations, followed by a presentation and discussion of the experimental results.

4.1 Setup

All measurements presented used the INA260 sensor connected in a high-side configuration. This configuration was chosen beacause the DUTs, especially the ESP32-C6, operate on a rather low current compared to the specifications of the sensor. Per configuration, platform and test case we performed 4 measurements of 4 s. Measurements were performed on two platforms:

BananaPi BPi-F3. This platform features a Spacemit K1 octa-core processor (up to 1.6 GHz) implementing the RV64GCVB/RVA22 instruction-set architecture. All experiments were executed on the vendor-provided Bianbu Linux distribution (version BPI-F3_Bianbu_23.10_v1.0.12_20240802). Non-essential system services were disabled, and the benchmark process was pinned to core 0 using the `SCHED_FIFO` scheduling policy. Despite extensive isolation efforts, we observed a recurring interference pattern: approximately every second, the supply current briefly dropped to approx. 0.28 A, close to the idle current, with a duration of approx. 45 ms. We were unable to fully eliminate this behaviour, but empirical observations indicate that the resulting disturbances do not materially affect the consistency of the measurements.

Espressif ESP32-C6. The second platform is the ESP32-C6 system-on-chip from Espressif Systems. It integrates both a high-performance (HP) 32-bit RISC-V core (up to 160 MHz) and a low-power (LP) 32-bit RISC-V core (up to 20 MHz). Both cores implement the RV32IMAC instruction set. All experiments were executed solely on the HP core. The on-chip Wi-Fi subsystem was disabled during measurement.

4.2 Test Cases and Configurations

To evaluate the system behaviour across a range of microarchitectural interactions, we designed 13 test cases (Table 3). Our goals were fourfold: (1) to evaluate the impact of state-flushing using very short test cases (`*_tiny`), where the relative influence of the flush should be highest; (2) to assess branch-prediction behaviour using randomized branch patterns (`branch_*`); (3) to exercise cache and prefetcher behaviour via diverse memory-access patterns; and (4) to study pipeline effects using test cases with predominantly independent operations or with tightly constrained dependency chains.

Table 3. Test cases executed for evaluation

Name	Functionality
`add_tiny`	Four additions on the same registers
`div_tiny`	Four divisions on the same registers
`read_tiny`	Four word-sized memory loads to different registers
`write_tiny`	Four word-sized memory stores from different registers
`branch_tiny`	Four not-taken conditional branches
`branch_hard`	Branches with a randomized taken/non-taken pattern
`cache_random`	32 loads and stores to random memory addresses
`cache_single_line_rw`	32 word-sized accesses within a single 64-byte cache line
`cache_stride_rw`	32 word-sized loads and stores with a 16-word stride
`chained_add`	128 additions to the same register (fully dependent chain)
`chained_div`	128 divisions to the same register (fully dependent chain)
`independent_add`	128 additions with read-write dependencies 4 instructions apart
`independent_div`	128 divisions with read-write dependencies 4 instructions apart

We evaluated the state-flush mechanism under four categories of configurations: (1) all methods enabled (`all_on`); (2) all methods disabled (`all_off`); (3) all but one method enabled, resulting in `no_`{`cache, bp, btb, pipe_drain, icache_sync`}; and (4) only a single method enabled, resulting in `only_`{`cache, bp, btb, pipe_drain, icache_sync`}.

To quantify the cost and variability introduced by each configuration, we executed both application benchmarks and a dedicated *empty* benchmark. The empty benchmark performs the full state-flush sequence and loop control but executes no operations, thereby isolating the intrinsic overhead of the flush routine itself. Using these measurements, we derive per-iteration energy and latency values, decomposing each run into a state-flush component and a benchmark-only component, as shown below. First, we compute per-iteration metrics for test case r as well as for the empty benchmark:

$$E_r^{\text{iter}} = \frac{E_r}{N_r}, \qquad t_r^{\text{iter}} = \frac{t_r}{N_r}$$

where E_r and t_r are the total measured energy and time of run r, and N_r is the number of iterations during the measurement. The energy and latency contributions of the pure workload of the test case, obtained by subtracting the state-flush baseline, are then

$$E_r^{\text{workload}} = E_r^{\text{iter}} - E_{\text{empty}}^{\text{iter}}, \qquad t_r^{\text{workload}} = t_r^{\text{iter}} - t_{\text{empty}}^{\text{iter}}.$$

4.3 Results and Discussion

Figure 4 shows the impact of the flush configuration on per-iteration energy and latency by means of a box plot. To directly quantify measurement noise, we also compute the coefficient of variation (CV), i.e., the standard deviation divided by the absolute value of the mean. A comparison of the CV across configurations is provided in Fig. 5.

Several effects are evident from the measurements. For energy consumption, two issues stand out. First, combining multiple flush methods substantially increases measurement noise. Second, some configurations introduce a systematic offset, resulting in baseline-subtracted values that even become negative. Both effects are more pronounced on the ESP32-C6 than on the BPi-F3. For latency, a similar pattern emerges, although in this case the disturbance is more severe on the BPi-F3. Overall, while disabling state-flush entirely yields the most stable results in most cases, a subset of benchmarks exhibit lower coefficients of variation under mild partial-flush configurations, suggesting that selectively applied state manipulation can still be advantageous.

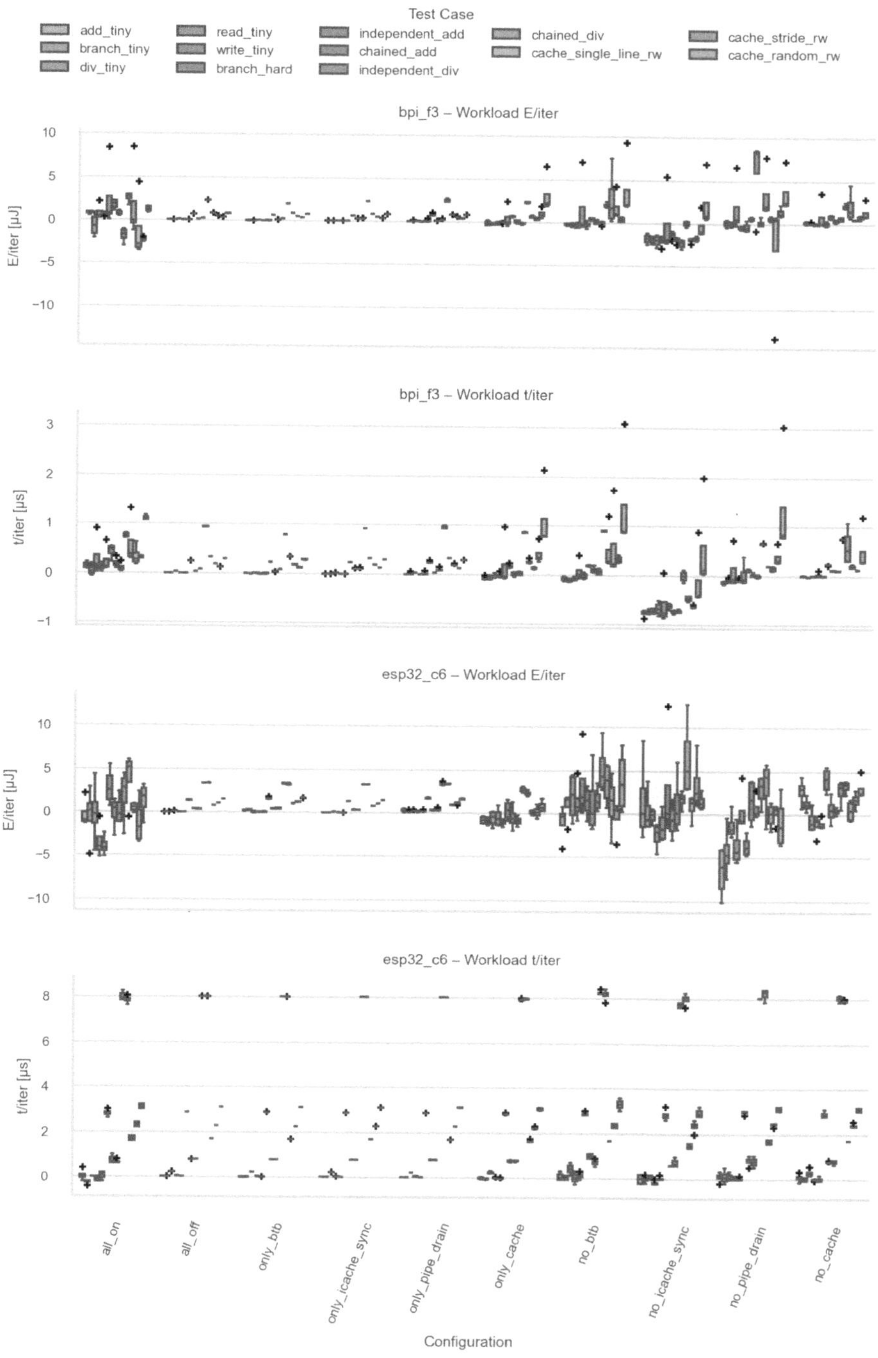

Fig. 4. Energy Consumption and Latency Distributions.

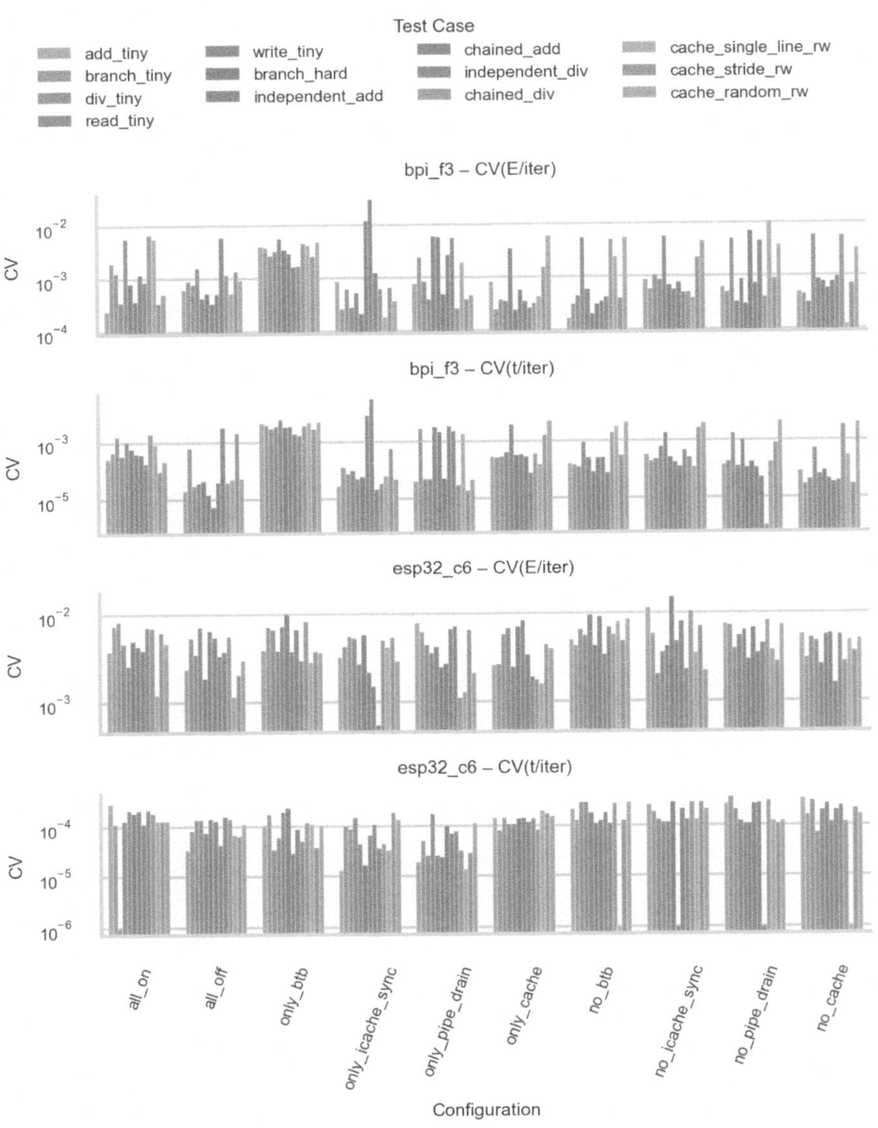

Fig. 5. Coefficient of Variation.

5 Conclusion

We introduced an open-source, automated framework that bridges system-level program execution with circuit-level measurements of energy consumption and latency. In contrast to prior approaches that focus primarily on instrumentation or measurement hardware, our framework emphasizes full end-to-end automation, encompassing program upload, configurable state-flush routines, controlled execution, and reproducible dataset creation. The current prototype, implemented entirely with low-cost COTS components and demonstrated on two different platforms, already provides robust and portable energy and latency characterization without requiring proprietary hardware support. Code and build instructions[1] are available to enable replication and extension by the community.

While the framework is fully usable in its current form, it remains work in progress. In particular, our evaluation revealed that the impact of different state-flush configurations is nuanced and, for some benchmarks, surprisingly strong. These effects warrant further investigation, ideally with larger numbers of repetitions per test case and across additional device families. Such an extended study would help distinguish inherent architectural behaviour from induced variance and would further strengthen the reliability of the derived per-iteration metrics.

We believe that this framework provides a practical foundation for systematic exploration of non-functional properties in embedded systems and will support future work on predictive modeling, compiler optimization, microarchitectural sensitivity analysis, and cross-platform comparison studies.

Acknowledgements. Artificial intelligence tools were employed for copy-editing purposes. The authors retain full responsibility for the scientific content. All ideas, conclusions, and data represent the authors' original contribution. We acknowledge support by the Federal Ministry for Economic Affairs and Climate Action (BMWK) due to an enactment of the German Bundestag under Grant No. 01MN23013H.

References

1. Chang, N., Kim, K., Lee, H.G.: Cycle-accurate energy measurement and characterization with a case study of the ARM7TDMI [microprocessors]. IEEE Trans. Very Large Scale Integr. (VLSI) Syst. **10**(2), 146–154 (2002). https://doi.org/10.1109/92.994992
2. Conti, F., et al.: Open source heterogeneous SoCs for artificial intelligence: the PULP platform experience. IEEE Solid-State Circuits Magaz. **17**(2), pp. 49–60 (2025). ISSN: 1943-0590. https://doi.org/10.1109/MSSC.2025.3537987
3. Dezfouli, B., Amirtharaj, I., (Chelsey) Li, C.-C.: EMPIOT: an energy measurement platform for wireless IoT devices. J. Network Comput. Appl. **121**, 135–148 (2018). ISSN: 10848045. https://doi.org/10.1016/j.jnca.2018.07.016

[1] https://gitlab.uni-rostock.de/twinspace/automated-measurement-setup.

4. Gomez, K., et al.: Energino: a hardware and software solution for energy consumption monitoring. In: 2012 10th International Symposium on Modeling and Optimization in Mobile, Ad Hoc and Wireless Networks (WiOpt), pp. 311–317 (2012)
5. Joulescope, Inc. Joulescope (2025). Accessed 18 Sept 2025. https://www.joulescope.com/
6. Lim, R., et al.: FlockLab: a testbed for distributed, synchronized tracing and profiling of wireless embedded systems. In: 2013 ACM/IEEE International Conference on Information Processing in Sensor Networks (IPSN), pp. 153–165 (2013). https://doi.org/10.1145/2461381.2461402
7. Nordic Semiconductor. Power Profiler Kit II (2025). Accessed 18 Sept 2025. https://www.nordicsemi.com/Products/Development-hardware/Power-Profiler-Kit-2
8. Qoitech AB. Otii Ace Pro. Datasheet/product specification (2024). Accessed 18 Sept 2025. https://www.qoitech.com/otii-ace/
9. Rusu, F.: CurrentRanger: auto-ranging nano-amp current meter (2018). Accessed 18 Sept 2025. https://hackaday.io/project/162509-currentranger
10. Shin, D., et al.: Energy-monitoring tool for low-power embedded programs. IEEE Design Test Comput. **19**(4), 7–17 (2002). https://doi.org/10.1109/MDT.2002.1018129
11. Al-Shorman, M.Y., Al-Kofahi, M.M., Al-Kofahi, O.M.: A practical microwattmeter for electrical energy measurement in programmable devices. Measur. Control **51**(9), 383–395 (2018). ISSN: 0020-2940. https://doi.org/10.1177/0020294018794350
12. Van Der Vlugt, S., et al.: PowerSensor3: a fast and accurate open source power measurement tool. In: 2025 IEEE International Symposium on Performance Analysis of Systems and Software (ISPASS). Ghent, Belgium, May 11, 2025, pp. 213–226 (2025). https://doi.org/10.1109/ISPASS64960.2025.00028

MultiGRA: Expanding the CGRA Design Space with a Mixed-Granularity Approach

Léo Pajot[1,2](✉), Quentin Dariol[2], Jérémie Crenne[2], Simon Rokicki[1], and Bertrand Le Gal[1]

[1] Univ Rennes, Inria, CNRS, IRISA, Rennes, France
{leo.pajot,simon.rokicki,bertrand.le-gal}@irisa.fr
[2] Keysom, Pessac, France
{qdariol,jcrenne}@keysom.io

Abstract. As computing needs are expanding, scaling up the general-purpose computing resources is limited by energy consumption and efficiency. Coarse-Grained Reconfigurable Arrays (CGRAs) are a solution to that challenge: they are versatile digital architectures, capable of achieving near-ASIC level of energy efficiency without sacrificing reconfigurability. However, CGRAs are limited to the instructions they were built with. Unsupported operations must be emulated, which severely impacts performance. To preserve CGRAs inherent features while enabling higher reprogrammability, we propose MultiGRA: an architecture template that expands the fixed CGRA's instruction set by including embedded FPGA fabrics (eFPGA) inside the CGRA as finely reconfigurable zones. The eFPGA areas are capable of implementing application-specific operators, specialized at compilation time. This approach avoids overprovisioning computing resources, since the architecture can be freely retargeted. We investigate two of the many integration possibilities of eFPGAs in CGRAs with the respective gains. Examples of post-synthesis ASIC area measurements are provided to show integration feasibility. In a case study comparing CGRAs and MultiGRAs, MultiGRAs show up to 40% improvement in execution time.

Keywords: CGRA · eFPGA · Heterogeneous PE · Reconfigurability

1 Introduction

Over the past decades, numerous hardware accelerator architectures have been proposed, each offering different trade-offs between raw computing performance, energy consumption, and implementation cost. While architectures deployed on Application-Specific Integrated Circuits (ASICs) and Field-Programmable Gate Arrays (FPGAs) deliver the highest performance and energy efficiency, they respectively lack flexibility and programmability. Conversely, General-Purpose Processors (GPPs) and GPUs offer high programmability but suffer from poor energy efficiency. A more balanced and flexible alternative lies in the design

M. Kowalczyk and C. Monière (Eds.): DASIP 2026, LNCS 16512, pp. 29–41, 2026.
https://doi.org/10.1007/978-3-032-23871-9_3

and use of CGRAs [16]. These grid-shaped architectures, which may resemble FPGAs or systolic arrays, trade FPGAs' bit-level flexibility against word-level configurability. Like GPUs, CGRAs are typically employed to accelerate specific portions of a program, offloaded from a host GPP. Related works [28] have demonstrated that CGRAs can achieve higher energy efficiency than GPUs and FPGAs for comparable performance levels, while requiring only a fraction of the hardware complexity.

This architectural simplicity comes at the cost of a more sophisticated compilation process, called mapping. It is a notoriously complex effort, as it depends on compile-time knowledge of the microarchitecture, but also on graph manipulations to solve for where on the grid, when, and in which sequence the instructions should be executed [17,18]. In this context, decreasing the number of instructions to be executed for the same workload may strongly impact mapping complexity and results, leading to shorter runtime cycle cost. Therefore, simplifying the kernel to be executed by fusing successive, simple operators together is key to improving execution-time performance, provided the Processing Elements (PEs) are able to execute such custom operations. To that end, PEs can be specialized by adding specific hardware operators in their available operation set. Adding application-specific hardware to a GPP has already been proven useful, for example in specialized RISC-V processors [2,3]. However, in this case, the tradeoff between application-domain specialization and retargetability arises, as an accelerator could be rendered obsolete if its Instruction Set Architecture (ISA) is too specific to an application. Special-purpose operators may also be less frequently used, making the added hardware cost difficult to justify.

In order to avoid those pitfalls while still augmenting CGRAs with domain-specific yet reconfigurable operators, we propose MultiGRA: a CGRA accelerator template using a mixed-granularity approach to application acceleration, not only implementing hot spots on a CGRA but also cycle-expensive sequences of operations to finely-reconfigurable zones inside that same grid. As discussed later, eFPGAs are a promising technology we investigated to implement such finely-reconfigurable zones. This multi-grained layout ensures that the strengths for which CGRAs are known are preserved as an extra degree of retargetable customization is enabled. The main contributions of this article are as follows:

- We propose a novel modular architecture that takes advantage of CGRA and eFPGA granularities and discuss its impact on extra hardware costs.
- We review opportunities to adapt a traditional CGRA workflow to achieve efficient compiling and mapping features while enabling operation fusion and their reconfigurable hardware implementation.
- We evaluate potential performance benefits achieved through integrating eFPGAs to PEs on some selected kernels.

The article is structured as follows: Sect. 2 covers background, while Sect. 3 presents the main distinctive features of the MultiGRA architecture template and workflow. Section 4 discusses hardware costs and performance benefits of our approach. Section 5 concludes and gives future work directions.

2 Related Works

2.1 CGRA Architecture and Frameworks

The concept of CGRA emerged in the early 90s, some preliminary implementations being described as "Explicit Data-Graph Execution" (EDGE) processors. Iterating on the first pioneer architectures like TRIPS [23] or the ADRES [18], current commercial and research CGRA accelerators are based on a mesh-grid architecture. Grid tiles, called *PEs*, manipulate data by performing word-level arithmetic or logic operations often referred to as *operators* in the CGRA literature. Each operator is implemented in a separate *Functional Unit* as shown in Fig. 1a. PEs also contain a private configuration memory that stores the execution program and compute/communication sequence of the tile. Tiles have direct connections with each neighbor, and a subset only has access to a shared data memory. Each cycle, every PE reads from their own configuration memory.

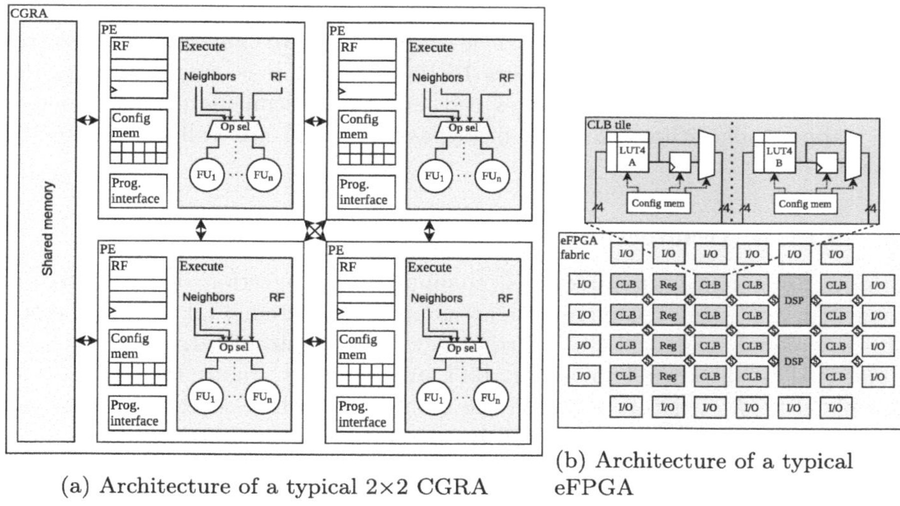

(a) Architecture of a typical 2×2 CGRA

(b) Architecture of a typical eFPGA

Fig. 1. CGRA and eFPGA architecture overviews.

The homogeneous architecture depicted in Fig. 1a is only an example, and many architectural variations were proposed, each with its own set of features and requiring a custom compilation toolchain to target that specific architecture. In order to help in designing and programming such systems, multiple open-source frameworks have emerged to model, generate Register Transfer Level (RTL) descriptions, simulate, and compile for CGRAs such as [11,22,25].

Programming a CGRA aims at identifying a computational kernel in the application source code and then efficiently mapping it onto the CGRA accelerator. This compile-time mapping process not only serves as a translation pass, bridging from the host processor ISA to the CGRA configuration words; it is also responsible for handling intra- and inter-loop data dependencies and managing routing resources selection. These are key steps in exploiting the architecture's computing performance. Different mapping approaches have proven to be effective to realize such tasks [6,17].

2.2 Specialized Operators

In CGRA and in GPP computing alike, a solution to reduce kernel execution time is to introduce application-dedicated operators in the architecture: in GPPs, this strategy has been applied in fields as diverse as Ultra-Low Power (ULP) signal processing [2,8], machine-learning transformer models acceleration [10], or cryptography [7]. In CGRAs, a notable example is the integration of costly floating-point operators [21], as well as other domain-specific operators [13].

However, adding dedicated instruction support involves implementing fixed resources, which may be underutilized based on the application, at the cost of surface usage and extra power consumption. Supporting these would also involve implementing multiple, underutilized operators in an overprovisioning fashion. To alleviate this issue, CGRA approaches already tried to chain simple operators together, creating complex operators to be executed in a single cycle [1,19]. Although effective, these techniques rely on very specific hardware resources and mapping algorithms to chain or bypass PEs and are still limited to the initial CGRA ISA.

2.3 eFPGA Architecture and Frameworks

eFPGAs are versatile architectures, enabling the integration of small, tightly integrated, and finely reconfigurable areas in larger systems. Like their discrete, chip-sized counterparts, eFPGAs are bit-level reconfigurable arrays of Look-Up Tables (LUTs) that can support any RTL architectures. Figure 1b illustrates an example eFPGA structure in which the number of IO blocks, the grid size, and the Configurable Logic Element (CLB) types may be modified. To help tailor these parameters to each use case, multiple open-source projects have emerged, enabling development of custom eFPGA fabrics [12,15,26].

Reconfigurable eFPGA architectures were investigated in System on Chip (SoC) system design to provide both flexibility and efficiency features. For instance, the silicon-proven FlexBex system combines a standard RISC-V core with a tightly coupled eFPGA [4]. Duet [14] also introduced a need for a tight integration of eFPGA zones in GPP datapath for fine-grained acceleration, in contrast to a GPP-FPGA association at the system level.

Work Motivation. The objective of tailoring CGRA architectures to a broad range of application domains extends beyond the general flexibility traditionally attributed to them. The promising results achieved in the FlexBex [4] and Duet [14] initiatives for GPP-centric systems motivated us to introduce eFPGA elements into CGRA architectures, with the aim of offering fine-grained optimization possibilities based on the application, while preserving flexibility through reprogrammability. In the following sections, we first present MultiGRA, a system that integrates FPGA resources within PE datapaths, and then evaluate the architectural costs associated with embedding eFPGA components.

3 Proposed CGRA-eFPGA System: MultiGRA

Previous work on CGRA focuses on homogeneous or heterogeneous grids that are fixed at the design time [9,20,24]. Different strategies for mapping computation kernels on these architectures have been studied, aiming at minimizing the execution time and maximizing tile utilization [17]. The MultiGRA concept targets the same objective, i.e., reducing kernel execution time on the CGRA architecture by introducing reprogrammable custom instructions, thus reducing the total number of instructions and execution time. To achieve this, we propose inserting eFPGA elements into the grid.

3.1 Overall Architecture

MultiGRA is designed to be loosely coupled with a GPP. Consequently, the CGRA is only responsible for executing a set portion of the program, determined at compile time. The data to process is streamed to the CGRA or stored in a shared memory accessible by a subset of PEs, as shown in Fig. 2. The novelty in the MultiCGRA approach consists in taking advantage of the concept of PE heterogeneity to examine possible eFPGA integration for flexibility purposes. Many works tackle PE heterogeneity in CGRA as an area- and energy-efficient way to distribute memory access or specific operators [1,9,25]. While heterogeneity makes for area-efficient accelerators, they lack post-fabrication flexibility, as the specialized PEs are not retargetable. We see this as an opportunity for eFPGA as a FU to limit the area overhead the eFPGA incurs and to avoid under-utilization. To that end, we propose and estimate the advantages and concerns of two distinct strategies to integrate eFPGAs in CGRA grids.

- As shown in Fig. 2, a first solution consists in integrating eFPGAs as FUs (Tile 3), alongside other operators implemented on a given PE. In that case, the eFPGA-augmented PE could execute both usual word-level operations, such as additions/subtractions or dataflow management operations, but also an arbitrary custom operator implemented on the eFPGA. This strategy extends the PE features but risks over-provisioning resources.
- A second solution, represented by Tile 4 in Fig. 2, favors implementing a larger eFPGA inside tiles, replacing entire PE computing resources. This approach replaces all the usual PE computing resources with the fabric and relies on the notion of PE heterogeneity. Due to being much larger than the first solution, the eFPGA fabric in this case can be used to implement more complex operators and whole circuits, which would not fit in a single FU.

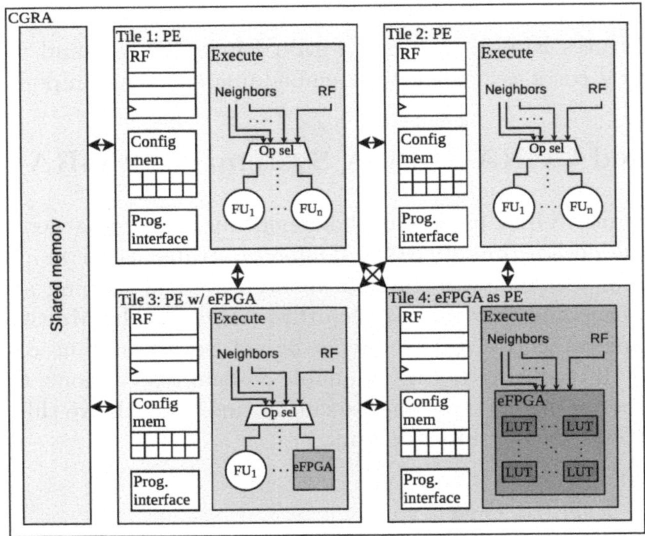

Fig. 2. Example of a MultiGRA architecture, highlighting potential eFPGA integration strategies.

Both approaches enable application-based customization of the CGRA. However, they incur different hardware overheads and enable various potential application speed-ups. In both cases in this study, we decided to focus on simpler eFPGAs only composed of CLBs, with no area-expensive resources such as DSPs. Furthermore, the eFPGA size would allow one or more custom operators, depending on their complexity. The benefits and drawbacks of both approaches on the hardware complexity and the mapping task are discussed during the experimental part of the paper.

3.2 Compilation Flow

The design methodology proposed to support the MultiGRA architecture is summarized in Fig. 3. The CGRA grid, eFPGA size, and integration strategy are design flow inputs according to tradeoff choices done by the designer.

First, the application described in C language is parsed and transformed (Fig. 3 ❶) into an Intermediate Representation (IR) format representing its Data Flow Graph (DFG). This stage is accomplished using Clang as currently implemented in the CGRA-Mapper tool [25]. The kernels to be accelerated on the CGRA are then manually selected among the identified candidates. The IR representation is then compared against a predefined list of potential custom operator patterns (Fig. 3 ❷). This pattern-matching process is performed under the eFPGA fabric's capacity constraint while searching for a legal set of custom operators with the highest performance improvement potential. Next, the original kernel IR is then modified, as identified patterns are replaced by a custom

IR node counterpart (Fig. 3 ❸). Note that the custom operator library and their associated IR patterns were previously designed such that their hardware complexity fit on eFPGA technology. Possible associations of RTL operators fitting the eFPGA tile's LUT budget are also described to ensure later deployment viability. The edited kernel IR operations are then scheduled on the augmented MultiGRA (Fig. 3 ❹), while the RTL models implementing the selected patterns are processed by an eFPGA synthesis and PaR tool (Fig. 3 ❺).

4 Evaluation

We evaluate the proposed approach by conducting an analytical assessment of the hardware complexity of MultiCGRA setups. Potential speed-up improvements of benchmark kernels on these architectures are also given to support our claims. The custom operation set for experimentation purposes comes from the standard ISA RISC-V Bitmanip extension [27] that was standardized according to its interest to speed up applications in multiple domains. The corresponding instruction patterns can be found in many benchmark applications, and the hardware cost in these RTL operators is low on eFPGA targets, as shown in Table 1a. This table presents the LUT cost of example Bitmanip operations on eFPGA technology produced respectively by the open-source frameworks FABulous [12] and PRGA [15]. These results are then used in the design methodology during the custom operation selection stage (Fig. 3 step ❷).

4.1 Hardware Cost Evaluation

To analyze the impact of eFPGA elements in a CGRA architecture, square PE grids were considered with sizes from 2×2 up to 6×6. The PE elements are homogeneous in the baseline architecture. The default PEs contain common word-level operators `add`, `shift`, `cmp`, `MAC`, `mul`, and `logic`, plus optional load-store operations, as well as control-flow support. The RTL architectures

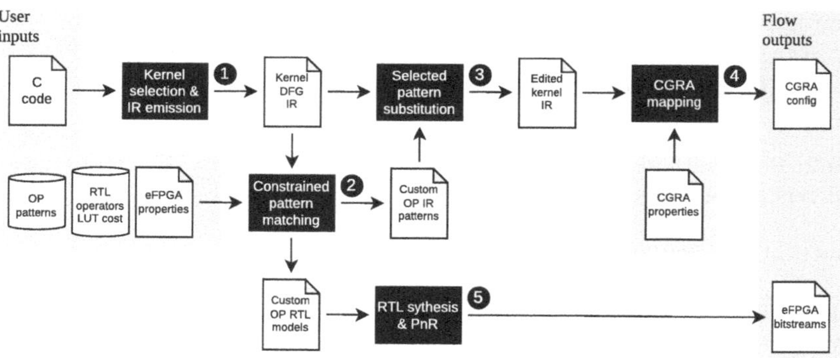

Fig. 3. Design methodology associated with proposed MultiGRA.

Table 1. Hardware cost estimations

(a) Post place and route LUT4 costs obtained using Yosys and NextPNR/VPR tools of some Bitmanip operators

Instruction mnemonic	FABulous (#LUT4)	PRGA (#LUT4)
`sh1add`	59	88
`andn`	32	32
`clz`	43	88
`max`	83	92
`rol/ror`	160	160
`bitmanip` (all instructions)	896	1 182

(b) Post-synthesis area results for homogeneous CGRA-Flow architectures according to the grid size

Target module	Area (μm^2)	Area (kGe)
CGRA grid (2×2)	59 094	120.7
CGRA grid (3×3)	137 769	281.4
CGRA grid (4×4)	249 850	510.3
CGRA grid (5×5)	395 243	807.3
CGRA grid (6×6)	573 977	1 172.3
CGRA Tile (full)	≈ 16 720	≈ 34.2
CGRA Tile (DataMem.)	≈ 700	≈ 1.4
CGRA Tile (FU)	≈ 7 903	≈ 16.1

were generated thanks to the open-source CGRA-Flow framework [25]. Table 1b summarizes the cost estimations obtained for these architectures using Synopsys Design Compiler under a 100 MHz clock frequency, targeting a commercial 28 nm CMOS ASIC standard-cell library. The silicon area costs of the CGRA key hierarchical components (1 complete CGRA Tile, 1 shared data memory, and 1 default FU set; see Fig. 2) are also reported. Ancillary elements, such as the configuration and routing hardware, are not reported for clarity. As expected, in these experiments, the hardware cost of CGRAs increases linearly with the CGRA grid size as PEs are homogeneous.

4.2 Reference eFPGA Study

To assess the overhead introduced by integrating eFPGA blocks into the processing elements, dedicated eFPGA modules were designed and characterized. To support at least two custom Bitmanip operations within a single eFPGA block, its size was set to 288 LUTs. For comparison, two RTL descriptions of eFPGA fabrics were generated using the FABulous [12] and PRGA [15] open-source frameworks, both configured to employ a frame-based programming scheme. The designs were then synthesized with Synopsys Design Compiler, targeting a 28 nm standard-cell technology under a 100 MHz clock constraint. eFPGA hardware costs obtained are summarized in Table 2.

Table 2. Post-synthesis hierarchical area estimates for two eFPGA fabrics generated by different frameworks

Framework	LUT4 count	Target clock (ns)	Area (μm^2)	Area (kGe)	LUT density ($LUT4/mm^2$)	Area per LUT4 (kGe/LUT4)
FABulous	288	100	175 762	359.0	1 638.6	1.246
PRGA			395 471	807.7	728.2	2.805

Reported values highlight the impact of the eFPGA generation framework on the resulting hardware cost. Although both fabrics implement the same total number of LUTs, significant architectural differences remain–most notably the LUT density per configurable logic block (CLB). FABulous natively integrates eight LUT4s per CLB, whereas PRGA includes only two. These design choices have major implications for fabric routability as well as LUT and CLB interconnect costs, as reflected by the observed 125% relative difference in area. Nevertheless, this study also demonstrates the robustness of both open-source frameworks, each producing working, synthesizable, parameterizable eFPGA architectures. It is worth noting that these area estimates were obtained using a generic cell library. As reported in [4], the area could be at least reduced by half when using a more specialized library and aggressive synthesis optimizations. Tables 1b and 2 indicate that, in the current configuration, a single default CGRA PE occupies an area equivalent to an eFPGA comprising 27 FABulous LUT4s. However, the results reported in [4] suggest that capacities on the order of one hundred LUTs may be attainable, potentially allowing most of the previously identified operators (Table 1a) to be supported.

Based on the presented results, we can now discuss the benefits and drawbacks of integrating an eFPGA as a FU (Tile 3, Fig. 1a) versus integrating an eFPGA as a PE (Tile 4, Fig. 1a). The first approach, which consists in embedding an eFPGA within the FUs of the PEs, would improve the overall utilization of PEs during mapping, since each PE could support both standard and custom operations. However, integrating an eFPGA with 30 to 100 LUTs would roughly double the CGRA hardware cost. To limit the global overhead at the CGRA level, such eFPGAs would have to be severely constrained in their LUT budget, which would significantly restrict the size and function of the custom operators they can implement to extend PE features. The second approach, which replaces a subset of PEs with eFPGAs, provides tighter control over the increase in CGRA hardware cost, thereby allowing a larger area budget for these blocks. This enables the implementation of more complex custom operators, but it also introduces a heterogeneous grid structure, which may lead to edge cases during mapping, as discussed in the following section.

4.3 Performance Gain Estimation

In order to validate our approach from a performance point of view and independently of eFPGA integration, this study compares three architectures. First, a reference CGRA (as generated by default by CGRA-Flow [25]). A custom operator is implemented in each of the second system's PE, alongside every other default FU (Tile 3, Fig. 1a). Only one custom operator is considered in the third system, replacing all the FUs of an arbitrary PE (Tile 4, Fig. 1a).

To obtain cycle counts of the CGRA with and without custom operators, a SystemC Transaction-Level Modeling (TLM) model was developed with the same structure as presented in [5]. As is default in CGRA-Flow, only the leftmost tiles of the grid have access to the data memory and thus implement load and store operations. Memory access penalties are not considered in this study, and data produced by a PE during a given cycle is considered available to its target neighbors at the beginning of the next cycle. Selected patterns were implemented by hand to model the corresponding eFPGA custom operator. CGRA-Flow's mapper (CGRA-Mapper) was also leveraged in order to produce kernel mappings for both reference and MultiGRA systems as described earlier.

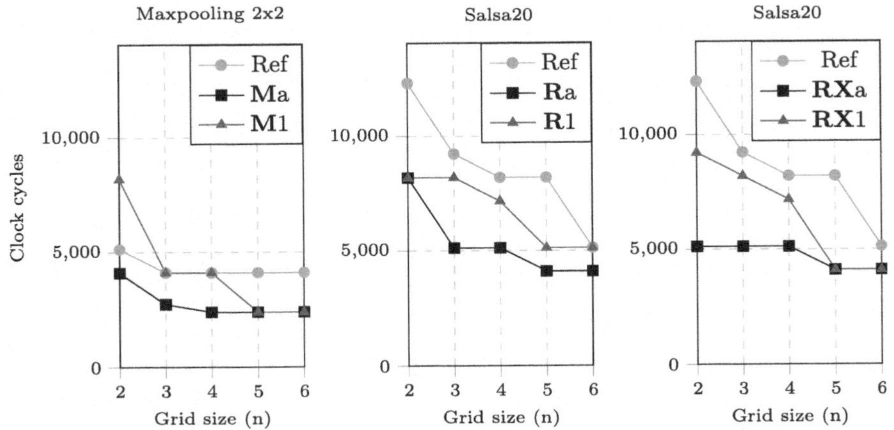

Fig. 4. Execution time of different kernels depending on CGRA and eFPGA configurations evaluated on 1024 kernel iterations.

In Fig. 4 a reference cycle count is compared to two custom operator implementation variants: the bold letter denotes the custom operator implemented (**M**ax is `maximum`(a,b), **R**OL is `rotate_left`(a,b), and **R**OL-**X**OR is `exclusive_or( rotate_left`(a,b), c)). Those three operations occur in many machine-learning, signal processing, or cryptographic workloads and are representative of the kind of operations found in the Bitmanip extension. An "a" or a "1" indicates whether the simulated architecture integrates the custom operator in all of its PEs alongside other FUs or in only 1 dedicated PE, respectively. For each tested kernel, the first CGRA system implementing the custom operator in all of its PE (**M**a, **R**a and **RX**a plots, black line with square markers) needs significantly fewer cycles than the same CGRA with no custom operator (Ref plot, orange line with circle markers) to execute the same number of iterations, reaching consistently between 30% and 50% relative cycle count reduction. This gap tends to reduce as the grid size increases, as more opportunities are available to the mapper to schedule interleaving kernel iterations. These results also highlight the impact the grid size has on performance, this trend being much more visible on more complex kernels, such as the cryptographic Salsa20 kernel.

The third set of lines (M1, R1, RX1, blue lines with triangle markers) models a system in which only 1 corresponding custom operator is implemented and in which the PE implementing it only supports that specific operator. Performance improvements are still achieved in many cases, up to 40%. However, the beneficial effect is reduced because of a side effect of specializing a given PE: the mapper is forced to relocate operations compared to the reference mapping solution, delaying the execution of subsequent operations and resulting in a longer execution time. This downside is more noticeable on smaller grids, resulting in a worse or on-par performance compared to the reference CGRA. However, in a MultiGRA system, this solution might still be worth considering, since dedicating full PEs allows for an area-efficient way to implement a larger eFPGA.

5 Conclusion

This work presents MultiGRA, a modular architecture closing the gap between usual CGRA grids and FPGAs by enabling custom operator implementation in eFPGA fabrics. These RTL-defined, application-specific, reconfigurable operators unlock domain-specific, post-fabrication specialization, improving on reference CGRA performance while maintaining cross-application retargetability at compile time. Many hardware design choices are available regarding eFPGA fabric insertion into the CGRA grid, varying in number and size. Using open-source frameworks, we explored eFPGA fabric integration costs, identified representative RTL operator LUT consumption, and demonstrated in system-level simulation cycle count reduction varying from 10% up to 40%. Our results highlight the impact of the target CGRA grid size on potential performance gains and pave the way to more in-depth discussions about intra-PE operator heterogeneity, eFPGA-augmented PE number and placement in the CGRA grid, and generalizing the operator substitution method to CGRA design space exploration.

References

1. Akbari, O., Kamal, M., Afzali-Kusha, A., et al.: PX-CGRA: polymorphic approximate coarse-grained reconfigurable architecture. In: Proceedings of Design, Automation & Test in Europe Conference & Exhibition (DATE) (2018)
2. Amor, H.B., Bernier, C., Prikryl, Z.: A RISC-V ISA extension for ultra-low power iot wireless signal processing. IEEE Trans. Comput. **71**(4) (2022)
3. Cui, E., Li, T., Wei, Q.: RISC-V instruction set architecture extensions: a survey. IEEE Access **11** (2023)
4. Dao, N., Attwood, A., Healy, B., Koch, D.: FlexBex: a RISC-V with a reconfigurable instruction extension. In: Proceedings of International Conference on Field-Programmable Technology (ICFPT) (2020)
5. Dariol, Q., Le Nours, S., Pillement, S., et al.: A hybrid performance prediction approach for fully-connected artificial neural networks on multi-core platforms. In: Proceedings of International Conference on Embedded Computer Systems: Architectures, Modeling and Simulation (SAMOS) (2022)
6. Dave, S., Balasubramanian, M., Shrivastava, A.: RAMP: resource-aware mapping for CGRAs. In: Proceedings of Annual Design Automation Conference (2018)
7. Fritzmann, T., Sigl, G., Sepulveda, J.: Extending the RISC-V instruction set for hardware acceleration of the post-quantum scheme LAC. In: Proceedings of Design, Automation & Test in Europe Conference & Exhibition (DATE) (2020)

8. Gautschi, M., Schiavone, P.D., Traber, A., et al.: Near-threshold RISC-V core with DSP extensions for scalable IoT endpoint devices. IEEE Trans. Very Large Scale Integrat. (VLSI) Syst. **25**(10) (2017)
9. Gobieski, G., Ghosh, S., Heule, M., et al.: RipTide: a programmable, energy-minimal dataflow compiler and architecture. In: Proceedings of International Symposium on Microarchitecture (MICRO) (2022)
10. Jiao, Q., Hu, W., Liu, F., et al.: RISC-VTF: RISC-V based extended instruction set for transformer. In: Proceedings of International Conference on Systems, Man, and Cybernetics (SMC) (2021)
11. Juneja, R., Dangi, P., Bandara, T.K., et al.: Building an open CGRA ecosystem for agile innovation (2025)
12. Koch, D., Dao, N., Healy, B., et al.: FABulous: an embedded FPGA framework. In: Proceedings of International Symposium on Field-Programmable Gate Arrays (2021)
13. Le, V.T.D., Pham, H.L., Tran, T.H., et al.: Efficient and high-speed CGRA accelerator for cryptographic applications. In: Proceedings of Eleventh International Symposium on Computing and Networking (CANDAR) (2023)
14. Li, A., Ning, A., Wentzlaff, D.: Duet: creating harmony between processors and embedded FPGAs. In: Proceedings of International Symposium on High-Performance Computer Architecture (HPCA) (2023)
15. Li, A., Wentzlaff, D.: PRGA: an open-source FPGA research and prototyping framework. In: Proceedings of Symposium on Field-Programmable Gate Arrays (2021)
16. Liu, L., et al.: A survey of coarse-grained reconfigurable architecture and design: taxonomy, challenges, and applications. ACM Comput. Surv. **52**(6) (2020)
17. Martin, K.J.M.: Twenty years of automated methods for mapping applications on CGRA. In: Proceedings of International Parallel and Distributed Processing Symposium Workshops (IPDPSW) (2022)
18. Mei, B., Berekovic, M., Mignolet, J.Y.: ADRES & DRESC: architecture and compiler for coarse-grain reconfigurable processors. In: Vassiliadis, S., Soudris, D. (eds.) Fine- and Coarse-Grain Reconfigurable Computing (2007)
19. Park, Y., Park, H., Mahlke, S.: CGRA express: accelerating execution using dynamic operation fusion. In: Proceedings of International Conference on Compilers, Architecture, and Synthesis for Embedded Systems (2009)
20. Prabhakar, R., Zhang, Y., Koeplinger, D., et al.: Plasticine: a reconfigurable architecture for parallel patterns. In: Proceedings of Annual International Symposium on Computer Architecture (2017)
21. Prasad, R., Das, S., Martin, K.J.M., et al.: Floating point CGRA based ultra-low power DSP accelerator. J. Sign. Process. Syst. **93**(10) (2021)
22. Ragheb, O., et al.: CGRA-ME 2.0: a research framework for next-generation CGRA architectures and CAD. In: Proceedings of International Parallel and Distributed Processing Symposium Workshops (IPDPSW) (2024)
23. Sankaralingam, K., et al.: TRIPS: a polymorphous architecture for exploiting ILP, TLP, and DLP. ACM Trans. Archit. Code Optim. (2004)
24. Swanson, S., Michelson, K., Schwerin, A., et al.: WaveScalar. In: Procedings of Digital Avionics Systems Conference (2003)
25. Tan, C., Agostini, N.B., Zhang, J., et al.: OpenCGRA: democratizing coarse-grained reconfigurable arrays. In: Proceedings of International Conference on Application-specific Systems, Architectures and Processors (ASAP) (2021)
26. Tang, X., Giacomin, E., Chauviere, B., et al.: OpenFPGA: an open-source framework for agile prototyping customizable FPGAs. IEEE Micro **40**(4) (2020)

27. Waterman, A., Asanović, K.: The RISC-V Instruction Set Manual, Volume I: User-Level ISA, Document Version 20191214-draft (2019)
28. Wijtvliet, M., Waeijen, L., Corporaal, H.: Coarse grained reconfigurable architectures in the past 25 years: overview and classification. In: Proceedings of International Conference on Embedded Computer Systems: Architectures, Modeling and Simulation (SAMOS) (2016)

Resource-Guided FPGA Floorplanning via Quad-Tree-Based NLP Optimization

Ze Chen(✉) and Jordi Cortadella

Universitat Politècnica de Catalunya, Barcelona, Spain
{ze.chen,jordi.cortadella}@upc.edu

Abstract. Field-programmable gate arrays (FPGAs) are widely adopted as accelerators for compute-intensive applications. In modern FPGA flows, floorplanning provides coarse-grained region constraints, but commercial tools offer no automatic support, requiring designers to manually define partition regions that substantially impact placement and routing. This paper presents a fully automatic framework that generates resource-guided floorplanning constraints to improve the clock frequency of FPGA designs. The framework employs a divide-and-conquer strategy with nonlinear optimization for resource-aware region assignment, exploiting FPGA heterogeneity in a device-agnostic manner and producing constraints that integrate into commercial toolflows. The framework has been integrated into Vivado and tested with nine benchmarks and two platforms, improving the maximum clock frequency by 11.4% and 5.8%, respectively, and yielding an additional 4.1% gain with timing-optimization strategies without significant congestion overhead.

Keywords: Field-programmable gate arrays · Floorplanning · Nonlinear optimization · maximum clock frequency

1 Introduction

Due to their reconfigurability, fine-grained parallelism, and energy efficiency, FPGAs have become key accelerators for performance-critical domains such as signal processing and machine learning [6,9]. In modern flows, designers rely on commercial placement and routing engines. However, these tools still depend on manually defined floorplans to guide physical synthesis. Effective floorplanning reduces design iterations and improves subsequent placement and routing.

Research on floorplanning spans multiple device generations. Early works introduced core representations such as slicing trees and sequence pairs, establishing the foundations for layout exploration via stochastic search [7,12]. Homogeneous FPGAs adopted similar formulations, enabling rectangle-based packing and annealing-based optimization [5,10]. With the advent of devices containing BRAM and DSP slices, floorplanning became a resource-aware allocation problem, with methods incorporating column alignment, multi-resource modeling, or analytical formulations to handle heterogeneous footprints [2,11,15].

M. Kowalczyk and C. Monière (Eds.): DASIP 2026, LNCS 16512, pp. 42–54, 2026.
https://doi.org/10.1007/978-3-032-23871-9_4

Still, a gap remains between academic formulations and modern industrial practice. Most prior work assumes rigid, non-overlapping rectangular regions with hard capacity constraints. In contrast, current implementation flows typically interpret floorplanning as soft region guidance: regions may partially overlap, and the placer is allowed to relax local suggestions when necessary to preserve routability or timing closure. At the same time, many real-world designs exhibit deep logical hierarchy and strongly heterogeneous resource requirements. These characteristics motivate formulations that explicitly account for hierarchy and heterogeneous resource footprints, while remaining compatible with soft region semantics and easy to integrate into industrial implementation flows. To the best of our knowledge, unfortunately, there is no literature on resource-aware floorplanning for single-die FPGAs with soft-occupancy constraints. To evaluate the impact of our contributions, we have enhanced a commercial flow with our soft-occupancy constraints and compared results with and without them.

The main contributions of this paper are summarized next:

- The problem of static floorplanning for heterogeneous single-die FPGAs under soft-occupancy constraints is formulated, bridging the gap between rigid academic models and practical industrial flows.
- A scalable divide-and-conquer approach is proposed, that combines hierarchy-aware netlist clustering with mathematical optimization to derive coarse-grained region assignments integrable into commercial CAD flows.
- The framework is architecture-agnostic and easily extensible, enabling deployment across diverse FPGA families and vendor toolchains.

2 Framework Overview

The framework is a pre-implementation stage between synthesis and placement. As illustrated in Fig. 1, it consists of four stages: (i) abstraction of the netlist and device layout, (ii) hierarchy-aware netlist clustering and quad-tree construction over the fabric, (iii) hierarchical floorplanning via a nonlinear programming (NLP) model, and (iv) post-processing to generate floorplanning constraints. This flow has been integrated into Vivado [14] to run experiments in Sect. 4.

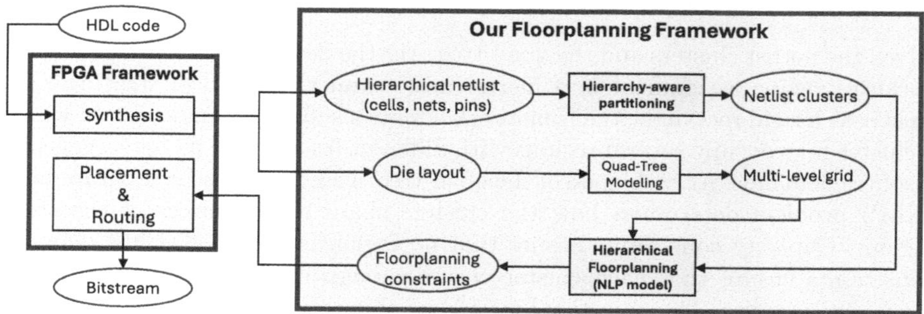

Fig. 1. Flow of the framework.

2.1 Hierarchical Netlist and Die Layout Abstraction

The starting point of the framework is a synthesized design and the description of the target FPGA architecture. From the synthesis output, we construct a netlist representation that captures cells, pins, nets and the hierarchical organization of cells into modules. In parallel, the device is abstracted as a two-dimensional arrangement of floorplannable units, each labeled with the types and relative amounts of available resources such as logic, memory, and DSP blocks. These units constitute the leaf level of a spatial hierarchy over the fabric. Reserved or forbidden areas of the device (such as configuration columns or dedicated clocking regions) are treated as non-allocatable regions. This abstraction provides a uniform architectural view for the subsequent clustering and floorplanning stages.

2.2 Hierarchy-Aware Partitioning and Quad-Tree Modeling

Given the abstracted netlist and device layout, the second stage builds two coupled hierarchies: a logical hierarchy of netlist clusters and a geometric hierarchy over the fabric.

On the netlist side, the framework performs a bottom-up, hierarchy-aware clustering to derive coarse-grained clusters with strong internal connectivity and limited external communication. Instead of flattening the entire design, the original module hierarchy is used as a guide, so that the clusters preserve module-level structural boundaries while reducing the problem size for optimization.

On the device side, the floorplannable units are recursively grouped into a quad-tree. Neighboring units are aggregated into larger regions, and this process continues until the whole die is represented by a single root node. Each node in the quad-tree represents a region of the heterogeneous fabric together with its available resources. The resulting multi-level grid provides a coarse-to-fine geometric decomposition that defines the search space for the subsequent region-assignment model. The detailed formulations of the clustering and quad-tree construction appear in Sect. 3.1.

2.3 Hierarchical Floorplanning with an NLP Model

Given the netlist clusters and the quad-tree over the device fabric, the third stage performs region assignment in a hierarchical manner. Instead of treating each cluster as a rigid rectangle, the framework adopts a soft-occupancy view in which a cluster may occupy several regions with different fractions of its heterogeneous resource demands. At each node of the quad-tree, a small nonlinear programming (NLP) problem determines how the clusters share the resources of the child regions. Capacity constraints ensure that no region is overfilled, while demand constraints ensure that each cluster covers enough resources across the child regions it occupies. The objective function combines two complementary effects: a connectivity-aware term, which encourages strongly connected clusters to stay

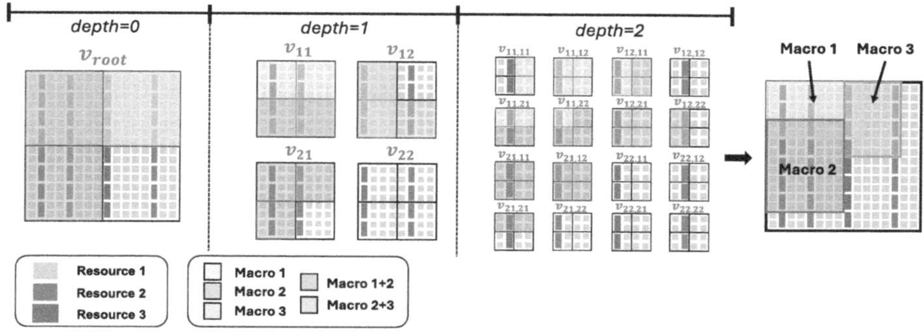

Fig. 2. Recursive refinement along the quad-tree hierarchy.

close, and a compactness term, which penalizes overly scattered allocations of a single cluster.

The optimization proceeds top-down along the quad-tree. Starting from the root, which captures a coarse global allocation, each level refines the distribution of occupancies within child regions. This multi-level scheme keeps individual NLP problems small, exploits the hierarchy of both design and device, and yields a set of soft region assignments that are later converted into floorplanning constraints. The full mathematical formulation of the NLP model is given in Sect. 3.2.

2.4 Floorplan Generation and Integration

When recursive refinement terminates, each cluster is associated with a set of regions where it has non-zero occupancy. Due to the soft-occupancy formulation, the corresponding footprint typically forms an irregular shape with staircase-like boundaries. Before emitting physical constraints, the framework encapsulates each cluster's occupied area in a minimum-area rectangular bounding box (Fig. 2), and overlapping boxes are allowed because the constraints are used as soft guidances. These boxes are finally translated into floorplanning constraints that can be consumed by commercial FPGA implementation tools.

3 Methodology

This section details the two main components of the proposed framework: hierarchy-aware clustering and quad-tree modeling of the FPGA fabric (Sect. 3.1), and the quad-tree–based NLP model for hierarchical floorplanning (Sect. 3.2).

3.1 Netlist Clustering and Fabric Abstraction

(A) Netlist-to-Cluster Modeling. Modern FPGA designs contain deeply hierarchical netlists with hundreds of thousands of primitive cells. The clustering stage recursively decomposes this hierarchy into coarse clusters with strong internal connectivity and limited external communication, in a bottom-up manner.

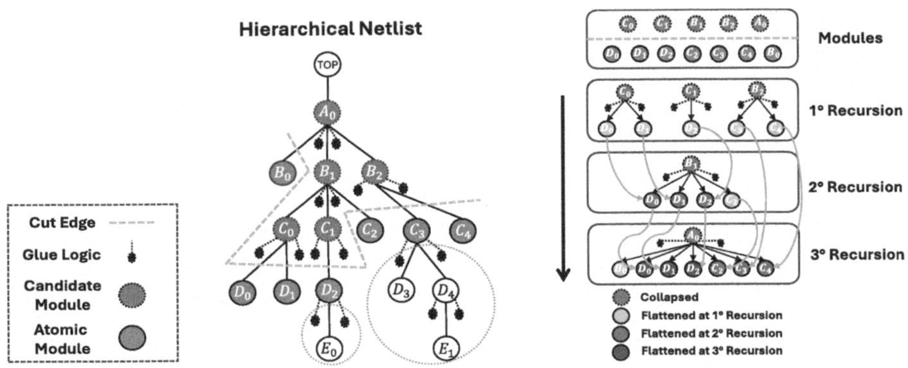

Fig. 3. Illustration of bottom-up recursive netlist partitioning.

As illustrated in Fig. 3, the netlist is first scanned to evaluate the aggregated resource demand of each module, i.e., each hierarchical logic block composed of multiple primitive cells. Modules whose resource usage exceeds a threshold are marked as clustering candidates, while smaller or leaf modules (with no children) are treated as atomic and permanently grouped with all their descendants.

We then traverse the hierarchy bottom-up and process candidates whose children are all atomic. For each such module, we build a local hypergraph capturing its glue logic and child connectivity, partition the glue logic among the children, collapse the parent, and reconnect the children directly to the parent's parent; this recursion continues until the design is flattened into a set of parallel coarse-grained clusters.

Formally, each design segment is modeled as a weighted hypergraph (V, E, ω, c), where each vertex v represents a primitive cell or module with resource demand $c(v)$, and each hyperedge e represents a net with weight $\omega(e)$ proportional to its bit-width. Hypergraph partitioning then produces clusters $C_1, \ldots, C_k$ by minimizing inter-cluster connectivity subject to balance constraints:

$$\min \sum_{e \in E} \omega(e) \cdot \delta(e); \qquad \text{subject to} \quad \frac{c(V_i)}{c(V)} \leq \frac{1}{k}(1+\epsilon), \quad \forall i \in \{1, \ldots, k\}, \tag{1}$$

Here $\delta(e)$ is the number of clusters spanned by hyperedge e, and ϵ is the imbalance tolerance. This hierarchical partitioning flattens the netlist into k coarse-grained clusters $\mathcal{M} = \{m_1, \ldots, m_k\}$, each with aggregated demand $\gamma_{n,t}$ over the type of resource t.

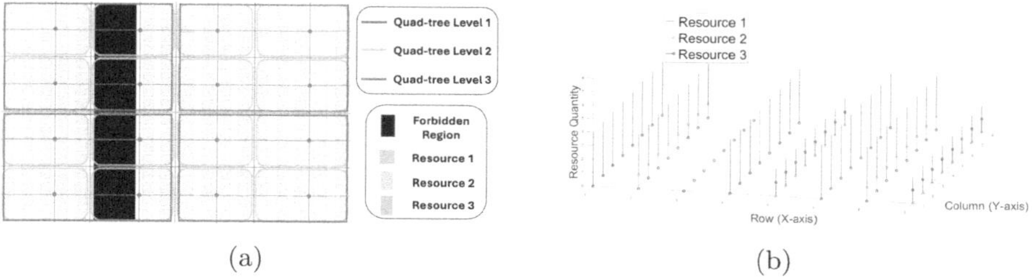

(a) (b)

Fig. 4. (a): Partitioning of heterogeneous FPGA layout (b): 3d footprint of heterogeneous resources of (a).

(B) Quad-Tree Abstraction of the FPGA Fabric. We now formalize the quad-tree construction sketched in Sect. 2.2. The heterogeneous FPGA fabric is aligned to a regular grid of base cells whose boundaries coincide with the smallest floorplannable units. Each base cell may contain multiple resource types, and its capacity vector counts how many resources of each type it contains. These cells form the leaf level of the spatial hierarchy. Figure 4(a) illustrates such a subdivision, while Fig. 4(b) highlights the heterogeneous resource distribution across base cells.

Starting from the leaf level, neighboring base cells are recursively grouped into 2×2 blocks to form parent nodes, and the process continues until the entire die is represented by the quad-tree root. Each node j is annotated with a capacity vector $\theta_{j,t}$ equal to the sum of its descendants. Forbidden or reserved regions (e.g., dedicated clocking or configuration columns) are included in the quad-tree by cutting along their vertical boundaries while aligning horizontal cuts with neighboring base cells, while the resulting cells have zero usable capacity.

The resulting quad-tree is the multi-level grid in which the NLP model allocates cluster resources in Sect. 3.2.

3.2 Quad-Tree-Based NLP Model

Building on the netlist clusters and the quad-tree grids, we now introduce the NLP formulation used for hierarchical node-level resource assignment. For clarity, the main symbols of the model are summarized in Table 1.

Each cluster m_n has heterogeneous resource requirements $\gamma_{n,t}$, and each quad-tree node j provides capacities $\theta_{j,t}$. The continuous occupancy variables $\rho_{j,n,t}$, which represent the fraction of m_n's type-t resource allocated to node j.

Table 1. Main notation used in the quad-tree-based NLP model.

Symbol	Description
m_n	Cluster n in the clustered netlist, $n \in \{1, \ldots, k\}$
j	Node (quadrant) in the quad-tree grids
t	Resource type (e.g., logic, memory, DSP)
$\gamma_{n,t}$	Demand of cluster m_n for resource type t
$\theta_{j,t}$	Available capacity of type t in node j
$\rho_{j,n,t}$	Fraction of m_n's type-t resource demand assigned to node j
(X_j, Y_j)	Geometric coordinates of the center of node j
(x_n^*, y_n^*)	Center of gravity of cluster m_n
$s_{j,t}$	Share of type-t resource capacity in node j
$\mathcal{R}$	Set of all nodes in the quad-tree
$\mathcal{J}$	Set of child nodes (quadrants) of the current node
E	Set of hyperedges in the clustered netlist
$\omega(e)$	Weight (bit-width) of hyperedge $e \in E$

Constraints. The first constraint explicitly bounds the occupancy variables:

$$0 \leq \rho_{j,n,t} \leq 1, \quad \forall j, n, t \tag{2}$$

The second constraint enforces the resource satisfaction for each cluster. Within the current node, the cumulative amount of type-t resources assigned to cluster m_n over all quadrants $\mathcal{J}$ must meet the demand $\gamma_{n,t}$:

$$\sum_{j \in \mathcal{J}} \rho_{j,n,t} \cdot \theta_{j,t} \geq \gamma_{n,t}, \quad \forall n, t \tag{3}$$

To couple allocations between heterogeneous resource types and satisfy the composition of each node, we normalize with respect to the aggregate capacity of that node. Let

$$s_{j,t} = \frac{\theta_{j,t}}{\sum_{t'} \theta_{j,t'}}$$

denote the share of type-t capacity in node j. The third constraint refers to the normalized cross–type usage bound. In quadrant j, the aggregate occupancy of resource t (summed over all clusters) must not exceed the normalized share $s_{j,t}$:

$$\sum_{n} \rho_{j,n,t} \leq s_{j,t}, \quad \forall j \in \mathcal{J},\, t \tag{4}$$

Connectivity-Aware Objective. The objective combines a connectivity-aware wirelength term with a compactness term. Since clusters are represented by fractional occupancies of quad-tree nodes rather than rigid shapes, the traditional half-perimeter wirelength (HPWL) is not directly applicable: there is no single bounding box per cluster, and geometry is defined implicitly by the $\rho_{j,n,t}$.

To estimate interconnection cost without requiring explicit rectangular footprints, we adopt a clique-based model that operates on clusters' center of gravity. For every net $e \in E$ connecting $n(e)$ clusters, a complete set of pairwise distances between cluster centers is formed, transforming the multi-pin connection into a fully connected graph (clique). The total interconnection cost is evaluated as

$$WL = \sum_{e \in E} \sum_{p=1}^{n(e)} \sum_{\substack{q=1 \\ q \neq p}}^{n(e)} d(x_p^*, x_q^*) \cdot \frac{2}{n(e)} \cdot \omega(e) \tag{5}$$

where $\omega(e)$ denotes the bit-width of hyperedge e, and $d(\cdot, \cdot)$ is the Manhattan distance between clusters' center of gravity. The L_1 norm can be smoothed, e.g., by replacing $|x_p^* - x_q^*|$ with $\sqrt{(x_p^* - x_q^*)^2 + \epsilon^2}$, to improve numerical robustness.

Centers of Gravity. Cluster's center of gravity (x_n^*, y_n^*) reflect the global spatial distribution induced by the occupancy variables. They are defined as capacity-weighted averages over all nodes $\mathcal{R}$:

$$x_n^* = \frac{\sum\limits_{j \in \mathcal{R}} \sum_t \theta_{j,t} \cdot \rho_{j,n,t} \cdot X_j}{\sum\limits_{j \in \mathcal{R}} \sum_t \theta_{j,t} \cdot \rho_{j,n,t}}, \quad y_n^* = \frac{\sum\limits_{j \in \mathcal{R}} \sum_t \theta_{j,t} \cdot \rho_{j,n,t} \cdot Y_j}{\sum\limits_{j \in \mathcal{R}} \sum_t \theta_{j,t} \cdot \rho_{j,n,t}} \tag{6}$$

During the hierarchical optimization, each NLP subproblem only treats the $\rho_{j,n,t}$ associated with the child nodes $j \in \mathcal{J}$ of the current quad-tree node as variables; all other occupancies over $j \in \mathcal{R} \setminus \mathcal{J}$ are inherited from previous levels and remain fixed. This preserves a globally consistent center of gravity while enabling localized refinement within each subproblem.

Dispersion Regularization. Minimizing Eq. (5) alone may lead to solutions where clusters are spread thinly across many nodes. To discourage such fragmented allocations, we introduce a dispersion term that measures the spatial coherence of each cluster. Dispersion quantifies how far a cluster's allocated fractions deviate from its own center of gravity:

$$D_{x,n} = \sum_j \sum_t \theta_{j,t} \cdot \sqrt{\rho_{j,n,t}} \cdot (x_n^* - X_j)^2; \quad D_{y,n} = \sum_j \sum_t \theta_{j,t} \cdot \sqrt{\rho_{j,n,t}} \cdot (y_n^* - Y_j)^2 \tag{7}$$

Since $\rho_{j,n,t} \in [0, 1]$, the transformation $\sqrt{\rho_{j,n,t}}$ amplifies the contribution of small fractional assignments, making it expensive for a cluster to occupy many nodes.

We aggregate the horizontal and vertical spreads into a single scalar dispersion measure by focusing on the dominant direction. A cluster is considered compact only if both its horizontal and vertical spreads are small; if one direction becomes much larger (e.g., an elongated stripe), the maximum of $(D_{x,n}, D_{y,n})$ reflects this lack of compactness. We therefore define

$$\text{Disp} = \sum_m \max(D_{x,n}, D_{y,n}) \approx \sum_m \frac{D_{x,n} + D_{y,n} + \sqrt{(D_{x,n} - D_{y,n})^2 + 1}}{2} \tag{8}$$

Final Objective and Hierarchical Optimization. The optimization problem at each quad-tree node is a nonlinear program that balances connectivity and dispersion:

$$\min_{\rho_{j,n,t}} \quad \alpha \cdot WL + \beta \cdot \text{Disp} \tag{9}$$

where α and β control the trade-off between wirelength and dispersion.

The NLP is solved recursively along the quad-tree, following the hierarchical scheme outlined in Sect. 2.3. Starting from the root node, which captures a coarse global allocation, each level refines the distribution of occupancies within its child quadrants while respecting allocations decided at higher levels. This multi-level strategy keeps each subproblem small, and ultimately produces the soft node assignments used to generate floorplanning constraints.

4 Experiments and Evaluation

Implementation and CAD flow. Post-synthesis netlist is exported from AMD Vivado 2024.2.2 and processed by a Python front-end, which builds the clustered netlist and quad-tree grids described in Sect. 3. Hierarchy-aware clustering uses the hypergraph partitioner *KaHyPar* [8], and each quad-tree subproblem is modeled in GEKKO [3] and solved with the IPOPT nonlinear optimizer [13]. The resulting per-cluster footprints are written as region constraints (pblocks) into a Vivado `.xdc` file which is either omitted (*baseline flow*) or sourced (*floorplan flow*) before batch implementation so that placement is guided by our framework.

Devices and Benchmarks. Experiments are conducted on two FPGA platforms, the AMD KRIA K26 and the Virtex-7 XC7VX415T. Nine benchmark designs are used for evaluation: four CNN accelerators generated by NN2FPGA [4], and five heterogeneous applications from the Koios suite [1]. These benchmarks span a wide range of sizes and resource compositions, including compute-intensive, memory-centric, and irregular workloads. The profiles are summarized in Table 2.

Vivado Synthesis Strategies. We evaluate both flows with two strategies [14]:

- **Default** strategy: the built-in balanced strategy, which aims to reach timing closure while maintaining moderate compilation time.
- **Performance_ExplorePostRoutePhysOpt(Perf_Opt)**: an aggressive timing-oriented strategy that applies multiple placement and routing algorithms and extensive post-place `phys_opt_design` to maximize timing slack.

Table 2. FPGA resource utilization for each benchmark design.

Benchmark/Device	CLB (count/%)	DSP (count/%)	BRAM (count/%)
KRIA K26	**14640**	**1248**	**144**
NA1	13005 (88.8%)	764 (61.2%)	51 (35.4%)
NA2	13181 (90.1%)	764 (61.2%)	46 (31.9%)
NA3	12480 (85.2%)	764 (61.2%)	43 (29.9%)
NA4	11415 (78.0%)	764 (61.2%)	57 (39.6%)
Virtex-7 VX415T	**64400**	**2160**	**880**
proxy6	10444 (16.2%)	99 (4.6%)	72 (8.2%)
proxy7	30254 (47.0%)	320 (14.8%)	352 (40.0%)
clstm	39379 (61.2%)	1289 (59.7%)	370 (42.0%)
bwave	14416 (22.4%)	700 (32.4%)	272 (30.9%)
dla	42469 (66.0%)	800 (37.0%)	96 (10.9%)

4.1 Methodology and Metrics

For each benchmark, implementation strategy, and flow (baseline vs. floorplan), we sweep the target clock period around the expected performance of the design. For each period T, Vivado is run to completion and the worst negative slack (WNS) is collected. The smallest period for which WNS ≥ 0 is denoted T_{min}. The corresponding maximum clock frequency is given by $f_{\text{max}} = 1000/T_{\text{min}}$ (in MHz). In addition to timing closure, we also record several metrics:

- **Routability.** Congestion information is obtained from Vivado's congestion reports. According to UG906 [14], congestion levels above 3 can degrade quality of result (QoR) and those above 6 often prevent timing closure or even lead to routing failure. Besides, Vivado typically terminates routing iterations early once timing closure is reached and further improvements are unlikely. We report the number of regions with congestion level > 3, and the number of global routing iterations.
- **P&R Runtime.** We report the runtime of the main implementation stages (placement, post-place `phys_opt_design`, and routing). The runtime of our pre-implementation framework is negligible and is omitted.

4.2 Results and Analysis

Figure 5 shows the maximum clock frequency for all benchmarks under both implementation strategies.

f_{max} **Under Default P&R Strategy.** Across all nine benchmarks, introducing pblock constraints improves the f_{max} compared to the baseline flow. For the CNN-based NN2FPGA designs on KRIA K26, the average improvement is **25.95 MHz (11.36%)**; for the Koios benchmarks on Virtex-7, the average gain

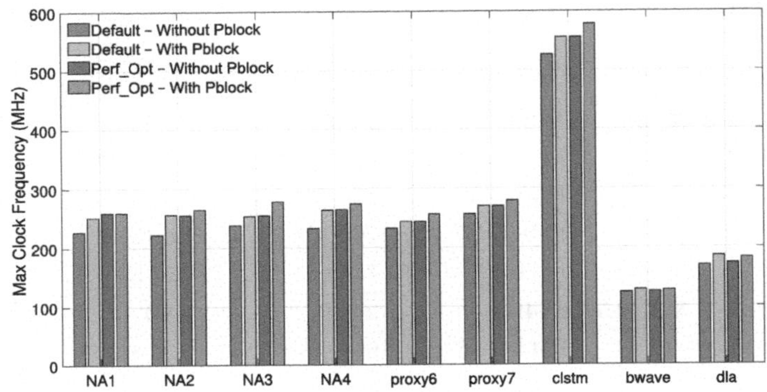

Fig. 5. Impact of pblock constraints on maximum clock frequency.

is **14.98 MHz (5.79%)**. In every design, the baseline flow fails to meet timing at the $f_{\max}$ achieved with pblocks, with worst negative slack deficits ranging from **0.047 ns** to **0.342 ns**. These results show that the proposed floorplanning consistently shifts the timing boundary to a more favorable operating point under a standard, non-aggressive implementation strategy.

$f_{\max}$ **Under Aggressive Timing Strategy.** With the Perf_Opt strategy, the baseline already achieves higher frequencies than with the default flow, and most benchmarks even exceed the pblock results obtained under the default strategy. Nevertheless, pblocks still push the frequency boundary further: on average we observe an additional average improvement of **10.93 MHz (4.06%)** over the corresponding baseline without pblocks. This indicates that global floorplanning and CAD-level timing optimizations are complementary: the former structures the placement space at a coarse granularity, while the latter refines critical paths within the assigned regions.

Congestion and Runtime. Table 3 summarizes two representative implementation points for each benchmark under the `Perf_Opt` strategy: (1) the run corresponding to the maximum clock frequency achievable without pblocks, and (2) the run corresponding to the maximum clock frequency achievable with pblocks. It provides a direct comparison of runtime and routability at the highest frequency each flow can achieve.

With pblock guidance, implementation runtime typically increases moderately, as placement, physical optimization, and routing explore a more structured, region-constrained search space to meet tighter timing objectives.

From the routability perspective, both congestion level and global routing iterations remain well behaved. For the larger NN2FPGA designs (NA1-NA4), which already exhibit non-negligible congestion in the baseline flow, the pblock-guided flow does not increase the number of congestion regions, and the number of global routing iterations is typically increased or unchanged, reflecting the additional effort required to meet tighter timing objectives in designs that

Table 3. Comparison of different runtimes (min), number of congestion regions (level > 3), and global routing iterations under **Perf_Opt** strategy with (pb) and without (npb) pblock constraints.

Benchmark	Placement		PhysOpt		Routing		Total		Cong. Reg.		GR Iter.	
	npb	pb	npb	pb	npb	pb	npb	pb	npb	pb	npb	pb
NA1	8:00	8:21	3:55	4:15	7:25	9:03	19:20	21:39	2	2	4	4
NA2	6:45	6:54	0:18	0:18	7:58	11:32	15:05	18:44	3	3	4	5
NA3	6:40	7:42	0:18	4:23	7:32	9:20	14:30	21:25	1	1	3	8
cre NA4	5:36	5:57	0:16	2:10	6:22	8:47	12:14	16:54	1	1	5	6
proxy6	2:10	2:20	0:24	0:28	3:51	5:00	6:25	7:48	0	0	4	3
proxy7	6:06	6:58	4:52	6:23	10:49	11:30	21:47	24:51	0	0	5	3
clstm	4:36	5:05	0:22	2:30	7:41	4:15	12:38	11:50	0	0	3	3
bwave	3:00	2:57	1:29	1:39	5:52	5:44	10:11	10:20	0	0	3	3
dla	4:02	4:13	0:15	0:14	3:22	3:26	7:39	7:53	0	0	2	2

are close to congestion limits. For the smaller Koios benchmarks, congestion levels remain stable, and the number of routing iterations is often reduced or unchanged, consistent with the intuition that, in relatively uncongested designs, coarse region guidance helps the router converge more quickly to a timing-closed solution.

Overall, under an aggressive timing-optimization strategy, pblock guidance achieves a higher frequency at the cost of a modest runtime increase, without introducing additional routing risk or significant implementation overhead.

5 Conclusion

We presented a resource-aware floorplanning framework for single-die FPGAs that integrates hierarchy-aware netlist partitioning with quad-tree–based NLP optimization. The approach can be incorporated into industrial flows. Experiments on nine benchmarks show consistent $f_{\max}$ improvements with modest P&R runtime overhead and without degrading routability, as reflected by stable congestion levels and similar routing-iteration behavior. Future work includes exploring shape-aware netlist modeling, extending the framework to multi-objective formulations that jointly consider timing, routability, and power, and applying the approach to emerging architectures with finer-grained heterogeneity.

Acknowledgements. This work has been supported by funds from the Spanish Agencia Estatal de Investigación under grant PID2020-112581GB-C21 (MOTION), and by the China Scholarship Council (CSC). The authors would also like to thank Luciano Lavagno and Yimamu Yilihamujiang for their helpful discussions and valuable feedback on this work.

Disclosure of Interests. The authors have no competing interests to declare that are relevant to the content of this article.

References

1. Arora, A., et al.: Koios 2.0: open-source deep learning benchmarks for FPGA architecture and CAD research. IEEE TCAD **42**(11), 3895–3909 (2023)
2. Banerjee, P., Sur-Kolay, S., Bishnu, A.: Floorplanning in modern FPGAs. In: 20th International Conference on VLSI Design held jointly with 6th International Conference on Embedded Systems (VLSID'07), pp. 893–898. IEEE (2007)
3. Beal, L., Hill, D., Martin, R., Hedengren, J.: GEKKO optimization suite. Processes **6**(8), 106 (2018)
4. Bosio, R., et al.: NN2FPGA: optimizing CNN inference on FPGAs with binary integer programming. IEEE TCAD **44**(5), 1807–1818 (2025)
5. Emmert, J.M., Bhatia, D.: A methodology for fast FPGA floorplanning. In: Proceedings of the 7th International Symposium on FPGAs, pp. 47–56 (1999)
6. Gandhare, S., Karthikeyan, B.: Survey on FPGA architecture and recent applications. In: Proceedings of the ViTECoN, pp. 1–4 (2019)
7. Murata, H., Fujiyoshi, K., Nakatake, S., Kajitani, Y.: VLSI module placement based on rectangle-packing by the sequence-pair. IEEE TCAD **15**(12), 1518–1524 (2002)
8. Schlag, S., Heuer, T., Gottesbüren, L., Akhremtsev, Y., Schulz, C., Sanders, P.: High-quality hypergraph partitioning. ACM J. Exp. Algorithmics **27**, 1–39 (2023)
9. Shawahna, A., Sait, S.M., El-Maleh, A.: FPGA-based accelerators of deep learning networks for learning and classification: a review. IEEE Access **7**, 7823–7859 (2018)
10. Shi, J., Bhatia, D.: Performance driven floorplanning for FPGA based designs. In: Proceedings of the 5th International Symposium on FPGAs, pp. 112–118 (1997)
11. Singhal, L., Bozorgzadeh, E.: Heterogeneous floorplanner for FPGA. In: 15th Annual IEEE Symposium on Field-Programmable Custom Computing Machines (FCCM 2007), pp. 311–312. IEEE (2007)
12. Wong, D., Liu, C.: A new algorithm for floorplan design. In: 23rd ACM/IEEE Design Automation Conference, pp. 101–107 (1986)
13. Wächter, A., Biegler, L.: On the implementation of an interior-point filter line-search algorithm for large-scale nonlinear programming. Math. Program. **106**, 25–57 (2006)
14. Xilinx Inc.: Vivado design suite user guide (2024)
15. Yuan, J., Dong, S., Hong, X., Wu, Y.: LFF algorithm for heterogeneous FPGA floorplanning. In: Proceedings of the 2005 Asia and South Pacific Design Automation Conference, pp. 1123–1126 (2005)

Sigma Delta DACs for Low Latency High Sampling Rate Audio on an FPGA

Jonas Höpner(✉), Romain Michon, and Anastasia Volkova

Inria, INSA Lyon, CITI, UR3720, 69621 Villeurbanne, France
kontakt@jonashoepner.de

Abstract. This paper presents a novel approach to ultra-low latency implementation of Sigma Delta (SD) Digital-to-Analog Converters (DACs) on Field Programmable Gate Arrays (FPGAs). We detail the design, simulations, and VHDL implementation of the proposed SD DACs. The paper provides an approach for a design-space exploration to find a tradeoff between the internal word lengths of calculations in the DAC and the Signal to Noise Ratio (SNR) of the output audio signal. In simulations, the proposed DAC can achieve an SNR of approximately 170 dB for input sample rates of 48 kHz - 768 kHz with the IP running at 49.152 MHz, resulting in an Oversampling Ratio (OSR) of 1024 - 64. In measurements, the DAC provides a latency inferior to 1 μs. Our SD DAC implementation has been integrated as an IP core to the SyFaLa toolchain, which aims to facilitate the prototyping of real-time audio Digital Signal Processing (DSP) applications on FPGA.

Keywords: DAC · FPGA · Sigma Delta · Audio Signal Processing

1 Introduction

Digital to Analog Converters (DACs) are at the heart of audio signal processing systems. For many applications that manipulate signals in real time, e.g., active noise control, it is important to achieve a latency in the microsecond range [4, 11,23,24]. One particular DAC technique, which became the industry leading standard for audio DSP, is Sigma-Delta (SD) DAC. It uses oversampling and noise shaping to convert digital signals into high-quality analog signals [3]. As Sigma Delta DACs are primarily digital, it is a natural candidate for Field-Programmable Gate Array (FPGA) implementations.

There are several good reasons for implementing an 1-bit SD DAC directly on an FPGA: (i) it eliminates the need for external audio codecs or costly hardware for additional audio outputs, (ii) this approach can help reach lower audio latencies and potentially high sampling rates (in the mega hertz range), which can be useful in the context of algorithms where aliasing is an issue.

Although Sigma-Delta modulators (SDMs) are not new, FPGAs started to have sufficient resources and speed to accommodate them in the late 1990s [5,12,14]. First works for 1-bit SD DACs for audio purposes were based on

M. Kowalczyk and C. Monière (Eds.): DASIP 2026, LNCS 16512, pp. 55–66, 2026.
https://doi.org/10.1007/978-3-032-23871-9_5

a reconfigurable third/fifth order cascade of integrators, feedback form (CIFB) for the modulator part. They implement an oversampling filter, the SD DAC and an analog lowpass after the DAC, reaching Signal-to-Noise ratios (SNRs) of 170, 140 and 138 dB.

In [10], a seventh-order SD DAC is implemented to generate a 1 kHz sine wave with 256x oversampling, so the signal has a sampling rate of 256 kHz. Using such a high OSR, we increase the SNR of the DAC further. However, it is not intended for audio processing but for ADC testing and does not oversample the 1 kHz signal with an oversampling filter, but by using a wavetable with the already oversampled data of 256 kHz sampling rate.

More recently, [13] implement a fourth-order SD DAC and an interpolation filter for oversampling, achieving an SNR of 139 dB.

All the aforementioned approaches employ a wavetable as the input for the SD DAC, which is straightforward to implement but occupies a significant portion of the FPGA's memory and is inherently restricted to reproducing only pre-programmed waveforms, making it unsuitable for non-predefined audio playback. Very few other sources discuss higher-order 1-bit DACs for FPGAs for audio applications. Several research papers [2,15] and open-source IPs[1,2] describe lower-order DACs.

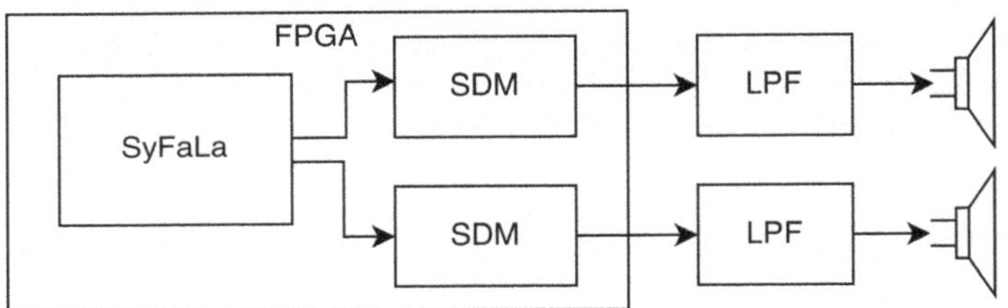

Fig. 1. Simple system diagram of the implementation of two 1-bit DACs (stereo output) in the context of the SyFaLa toolchain and the corresponding analog lowpass filters for signal reconstruction. Because of the high sampling rate being used, the order of these filters can remain relatively low and a simple one pole filter can do the job.

This preliminary work aims to bridge the gaps between on-chip high-order low-latency SD DAC implementations on FPGAs.

We propose a new implementation of 1-bit DACs with SDMs up to fifth order, integrated into the Syfala toolchain[3] [18] (see Fig. 1). Syfala facilitates the design of real-time audio DSP applications by allowing programmers to write high-level DSP code in C++ or in the Faust programming language [17]. Syfala takes care of bridging audio DSP IPs with audio codecs, potential sensors for control, etc.

[1] https://github.com/hamsternz/second_order_sigma_delta_DAC.

[2] https://github.com/davemuscle/sigma_delta_converters.

[3] https://inria-emeraude.github.io/syfala/.

The DACs described in this paper provide an audio latency inferior to 1 μs, from input to output, which corresponds to approximately 20-50 clock cycles at clock speeds of around 49 MHz, making them suitable for active noise control applications on FPGAs [1]. At 49 MHz, our DAC achieves a maximum SNR of 170 dB, demonstrating superiority wrt existing approaches. The VHDL implementation, integrated into the SyFaLa HLS tool is open-source.[4]

To optimise the balance between DAC size and speed, a pareto front of bit-perfect output was created, comparing an implementation using 64 bit Floating Point (FP) calculations as a baseline to an implementation of Fixed Point (FxP) of varying lengths to find a minimum size of the algorithm on the FPGA which still performs the same. Additionally, with the same simulation framework, a minimum size for a given target SNR was searched, which then, due to the smaller size of the implementation, could reach higher speeds on the FPGA. The smaller implementation can partly offset the lost precision due to its higher achievable OSR.

With an SD DAC, two clocks are differentiated: First, the audio Sampling Rate (SR), which describes the rate at which new samples of the input audio arrive. For audio applications, the SR usually lies between 44.1 kHz and 768 kHz. The second clock is the speed of the Sigma Delta Modulator (SDM) itself, so the speed of the SD algorithm on the FPGA. For each clock cycle, the algorithm generates one bit on the output General Purpose Input/Output (GPIO). This is a one-bit, high-SR output. As the SDM usually runs many times faster than the input signal, it uses an upsampled signal, usually by 64-256 times, which then results in OSRs of 64 to 256. As explained earlier, higher OSRs push the quantisation noise further up, increasing the SNR of the system. Our fifth-order DAC is able to run at around 49 MHz, enabling OSRs of up to 1024 for 48 kHz input sampling rate, therefore offering a higher SNR compared to slower configurations. Similarly, this enables input sampling rates of up to 768 kHz with an OSR of 64 for a higher resolution/bandwidth of the input signal. As the state of the art runs slower than our DAC, this project is the first to support these high output- and input SR. As an example: Considering a standard audio input rate of 48 kHz, and DAC clock of 49 MHz: a very high OSR of 1024 is achieved, which results in improvements to the SNR, as the quantisation noise is pushed to even higher frequencies. Having the quantisation noise in higher frequencies eases the requirements to the lowpass filter (i.e., order, etc.), as noise in higher frequencies is filtered out stronger.

With an SD DAC on-chip, this project proposes an alternative to audio codecs on development platforms around FPGA or external codecs. This is also an advantage for projects which require a high number of output channels, e.g. spatial audio. Spatial audio profits from a possible high channel count, as each channel can be used as a audio source of a different direction. Utilising an FPGA to create spatial sounds with a high number of speakers (usually a low to mid two-digit number) positioned throughout a room can be achieved using this

[4] The code is at https://github.com/inria-emeraude/syfala/tree/main/source/rtl.

project's direct FPGA output, eliminating the need for costly hardware to provide additional outputs.

This project is carried out in the context of SyFaLa and the Faust programming language. Faust is a Domain Specific Language (DSL) targeting digital audio processing and synthesis. SyFala is a framework to automatically compile Faust and C++ DSP programmes for Xilinx FPGA. SyFaLa offers automatic hardware implementation of Faust and C++ programmes, and also is responsible for the management of the software side on the FPGA. Concerning this project for example, SyFaLa manages the Faust programme and the layout of the IP on the FPGA, automatically instantiating the needed number of DAC.

2 Proposed Approach

2.1 The Sigma Delta Loop

Sigma Delta Modulation is a method for converting high-bit, low-sample-rate signals (e.g., 16-bit, 44.1 kHz) into low-bit, high-sample-rate signals (e.g., 1-bit, 2.8 MHz) for digital-to-analog, analog-to-digital and analog-to-analog conversion. To account for the quantisation noise emanating from the shortening of the conversion to lower bit depths, SDM oversamples the signal, then continuously corrects and shifts the quantisation noise to high frequencies using a negative-feedback loop [3], i.e., performing "noise shaping." Oversampling increases the sampling rate higher than required by Nyquist, thereby distributing the emanating quantisation noise over a broader frequency spectrum. For example, if a signal is sampled twice as fast as required by Nyquist, its OSR is 2.

A simple analog lowpass filter can then be employed to filter out the noise, making the process relatively inexpensive in terms of required hardware. These filters can be built easily, e.g. using a capacitor and a resistor for a first-order low pass filter. For the aims of this project, this is an advantage against a multi-bit SDM and following DEM, as this would necessitate more parts after the FPGA, instead of the simple design consisting only of the FPGA and a low pass filter.

When sampling a system at the Nyquist rate with *high bit depths*, the system directly benefits from the accuracy of the signal's amplitude, but it requires highly precise equipment to maintain linearity. Sigma Delta DACs use a *low bit depth* but seek higher accuracy in the time domain. They achieve high linearity and impose fewer requirements on the lowpass filter after the DAC or before the Analog-to-Digital Converter (ADC) [16].

To go into more detail on how SDMs works, consider the concept of digital density. For instance, to achieve 75% of a reference voltage, one could either use a circuit that directly provides 75% of the voltage or one that provides 100% of the reference voltage for 75% of the time. Both methods average to 75% of the reference voltage.[5] A comparable example is Pulse Width Modulation (PWM) in display screens, where brightness control is achieved by rapidly switching the display on and off. By lowering the brightness, the screen flickers rapidly,

[5] https://www.ieee.li/pdf/essay/calculating_delta-sigma_snr.pdf.

providing full brightness for shorter periods. To the human eye, this appears as a lower overall brightness. Adjusting the PWM duty cycle changes the perceived brightness. SD employs Pulse Density Modulation (PDM), which modulates a signal using pulse density. PWM is a type of PDM where all pulses for one sample are grouped together in the digital signal. Both PWM and PDM signals can be converted to analog signals using a lowpass filter. For example, with a resolution of 4 bits, a PWM waveform representing a 50% signal will be on for 8 clock cycles and off for the remaining 8 cycles. In contrast, a PDM signal at the same clock rate will switch every cycle. When passed through a lowpass filter, both waveforms produce an average signal level of 50% of the maximum. The output of a 1-bit SDM is essentially a Pulse Density Modulation (PDM) signal encoding an average voltage value.

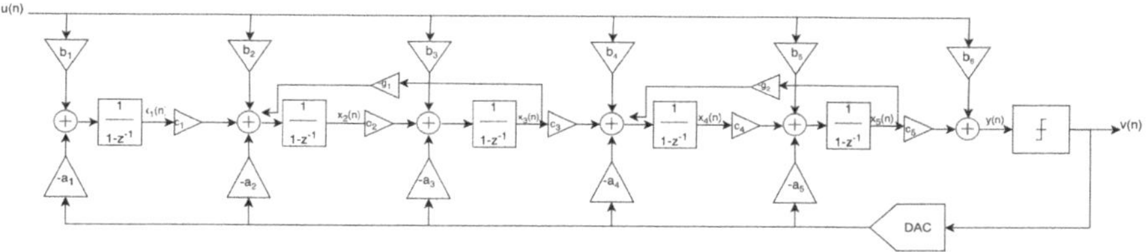

Fig. 2. A model of the implementation of the fifth-order CIFB DAC.

2.2 Topologies of SD DACs

Figure 2 illustrates a fifth order SD DAC, excluding the upsampling and lowpass filters. The SD process begins with the Delta stage, where the oversampled input, multiplied by (b_n) and negative feedback $(-a_n)$ are combined. It then proceeds to the Sigma stage, which functions as an integrator, integrating all its inputs. The number of integrators determines the "order" of an SDM, here equal to five. As the output at the comparator of a 1-bit SD DAC is binary (1 or 0), the depicted internal DAC for the feedback path can be substituted by a multiplexer for the sign of the coefficients a.

At the end of a standard SDM is a quantiser; in a 1-bit SDM, a comparator suffices. In this project, the comparator output is sent to the GPIO of the FPGA, which acts as the DAC: output bit 1 sets the GPIO high, and output bit 0 sets it low. This output is a low-bit, high-SR signal, also known as the bitstream. To convert this signal to analog form, an analog lowpass filter is used to remove high-frequency components, making the signal time- and value-continuous. Additionally, SDMs can achieve "analog-to-analog" conversion, recording the bitstream for later playback, as used in Direct-Stream-Digital (DSD) audio players.

The coefficients a determine the poles of the Noise Transfer Function (NTF) and Signal Transfer Function (STF), coefficients b are responsible for the zeros of the STF, and coefficients c are used for dynamic range scaling (in our approach, all coefficients `c` are set to 1). To achieve a unity STF, the coefficients `a` and `b` must be equal, as the DC gain of the STF is b_n/a_n [19].

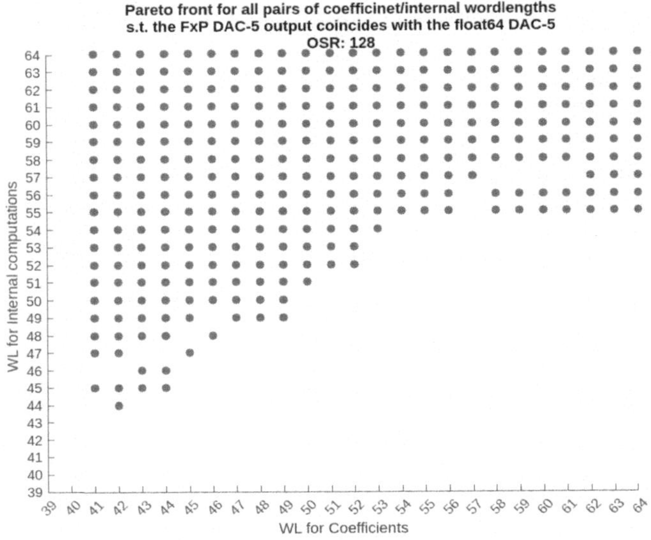

Fig. 3. The Pareto front of the WL assignment for DAC coefficients vs. internal rounding, while matching of the output of a real-valued (float64) corresponding DAC.

2.3 Fixed-Point Design-Space Exploration

A straightforward approach is to use floating-point arithmetic for the SDM implementation but it requires a significant amount of FPGA resources and it compromises the maximum possible clock of the IP.

A more efficient way is to use integer-based fixed-point computations. The primary challenge in fixed-point circuit design is: (i) the determination of correct ranges for all intermediate signals and (ii) the allocation of enough precision for intermediate computations to obtain an acceptable level of accuracy.

This Design-Space Exploration (DSE) is a difficult combinatorial problem, especially if a model of the error propagation through the circuit is available. As this paper presents a preliminary work, we have first designed a simulation framework in Matlab, which varies two parameters: internal wordlengths (WLs) and transfer function coefficient (`a` and `b`) sizes. When referring to an `(N/M)` configuration, we simulate a rounding to `N` bits after each arithmetic operator and rounding the filter coefficients to `M` bits.

As a first approach, we propose to take as a quality baseline a double-precision floating-point implementation and verify if the final comparator (quantizer in

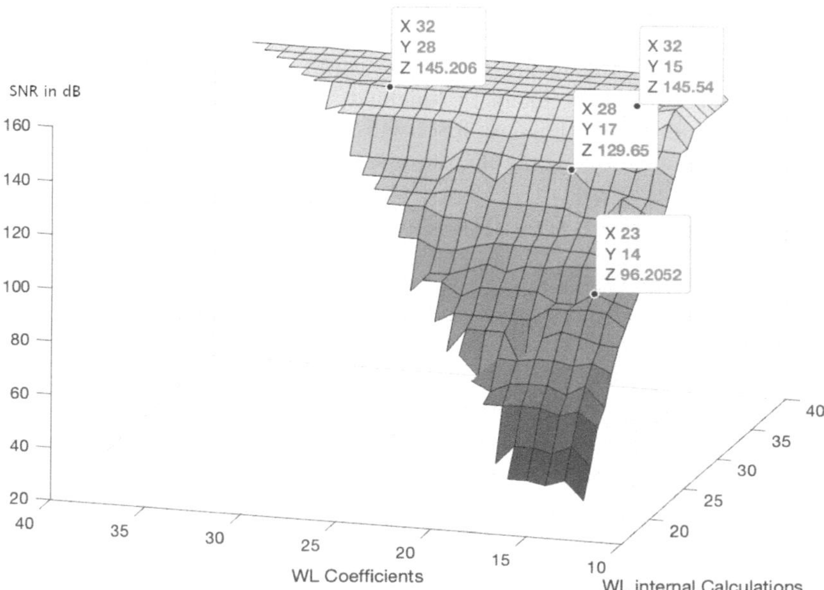

Fig. 4. Tradeoff between SNR, the WLs of internal calculations and of coefficients. The SR is 48 kHz, and an OSR of 128.

Fig. 2) is equivalent between the low-precision fixed-point version and the baseline. See Fig. 3 for the resulting Pareto front, indicating that the minimal configuration maintaining the double-precision quality is (44/42). However, as it will be detailed in the next section, practical implementation of such a configuration may not be feasible for small and mid-size boards due to a significant amount of resources needed. Obviously, reducing the WLs introduces quantisation errors and decreases the Signal to Noise Ratio (SNR). As a fifth-order 64-bit FP SDM can achieve over 200 dB SNR, the lower performance of shorter WLs is acceptable for audio applications.

Hence, we incorporate the SNR evaluation into the simulation framework. We fully parametrise the simulation by the desired OSR. The simulated output is processed through a lowpass filter with the following settings: cutoff frequency of 24 kHz, stopband attenuation of -150 dB, and a steepness of $1 - 10^{-12}$. Then, the SNR of the signal is computed.

Two primary factors influence the DAC's performance: the WLs of internal calculations and coefficients, and the input sample bit depth. Moreover, the input WL also impacts the usage of DSPs for multiplications, depending on the board vendor (25 bits for the Xilinx DSPs). These parameters are configurable in our simulation framework and permit, for example, to establish Pareto fronts as in Fig. 4.

Note that these simulations represent theoretical maximums. For example, the theoretical maximum SNR for a perfect fifth-order DAC with 1024x OSR is 281.74 dB. Achieving even the simulated 170 dB in a real-world scenario is challenging, as the following lowpass filter would need to achieve at least the same SNR, and wires must likely be shielded. In real-life implementation, the thermal noise of an analog lowpass filter will further limit the system's SNR.

The DSE framework is available and open-source[6].

3 Implementation Results and Discussion

We have applied our design-space exploration routine to SD DACs of order 1 through 5 and implemented them on a Zybo Z7-20 development board. The Zybo Z7-20 is a development platform based on the Xilinx Zynq-7000 series XC7Z020-1CLG400C SoC, manufactured by Digilent. It features 13,300 logic slices, 53,200 slice LUTs, 220 DSP slices, and 106,400 flip-flops, operating typically at 125 MHz.

The output GPIO of the DAC on the FPGA is managed by SyFaLa, making it possible to create many DACs without manual configuration. For example, SyFaLa would instantiate four DACs for the following Faust program (Fig. 5):

```
import("stdfaust.lib");
process = par(i,4,os.sawtooth(100 + i*100));
```

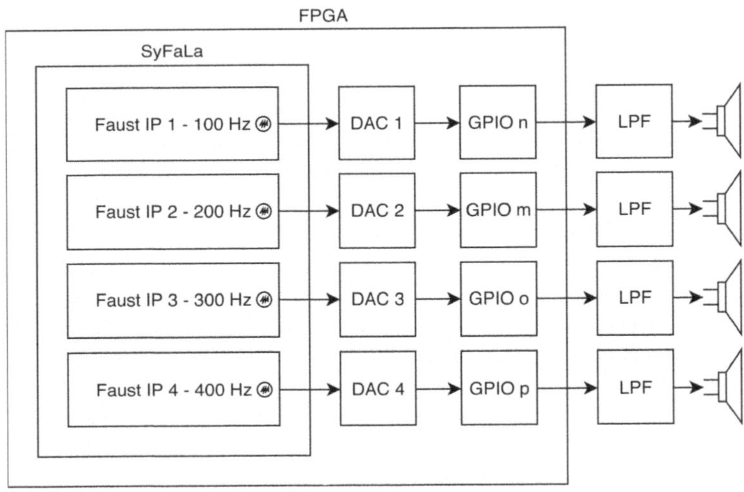

Fig. 5. An implementation of 4 sawtooth generators with Faust.

This program puts four sawtooth waves in parallel, with different frequencies of 100, 200, 300 400 Hz respectively. The outputs of the DACs would then each be mapped to a GPIO pin of the FPGA.

[6] The code is available at https://github.com/jhoepner/sigma-delta-matlab.

Fixed-Point Design. The DAC's coefficients were generated using the DSToolbox.[7] In the SD loop, the feedback value is multiplied by the DAC's output, resulting in either a_n or $-a_n$. To remove the multiplications, a multiplexer sets the coefficient's sign based on the DAC's output, using only addition in the loop. Internal calculations employ truncation over rounding, potentially reducing accuracy but enabling higher speeds and increased OSR, which pushes quantisation noise to higher frequencies, simplifying post-filtering.

DSE and Resource Consumption. We apply our DSE framework for SD DACs of orders 1 to 5 to find the Pareto-optimal points, taking into account the maximum sizes of the DSP slices (here 25 by 18 bits). Table 1 presents the IP size with their internal and coefficient WLs. The optimal fifth order DAC with a `(34/17)` configuration was able to run at OSR 1024 at 49.152 MHz, achieving 170 dB SNR.

The first-order DAC operates at 125 MHz, matching the FPGA's clock speed, with no observed speed limitations. However, higher-order DACs are constrained by speed limitations and could not run at the full FPGA clock rate. Their respective maximum operating speeds are listed in the table.

Notably, the third-order DAC using FP arithmetic exhibits substantial resource demands. Similarly, the fixed-point third-order DAC requires extensive resources due to its long coefficient word lengths, occupying more space than even the fifth-order DAC with comparable internal word lengths. The final entry – a fifth-order DAC with a 37-bit internal word length – was unable to generate samples in time, necessitating a reduced clock speed for the DAC. Consequently, this configuration is unsuitable for an OSR of 1024 at a 48 kHz input SR and will not be considered further.

Vivado reported an energy consumption of 3 mW for the fifth-order DAC in a 24-bit input configuration `(34/17)`, using 329 slices, 146 slice registers, and

Table 1. Resource Usage of Different SD DAC Orders and Approaches. SNR values are the simulated maximums (SR 48kHz, OSR 128).

Order	Clock Freq. (MHz)	Internal WL(bits)	Slice LUTs	Slice Registers	# DSP	SNR (dB)
1 (int)	125	19	21	37	0	56
1 (FxP)	125	19	47	36	0	56
2 (FxP)	41.66	27/17	346	77	3	81
3 (FP)	<1.25	float16	8652	66	4	105
3 (FxP)	38	35/35	423	139	7	105
4 (FxP)	28	27/17	425	106	3	125
5 (FxP)	52	27/17	281	116	3	117
5 (FxP)	52	34/17	329	146	3	144
5 (FxP)	<52	37/17	349	155	3	144

[7] https://www.mathworks.com/matlabcentral/fileexchange/19-delta-sigma-toolbox.

3 DSP slices. This accounts for less than 1% of the FPGA resources, leaving space for additional IPs. As a result, this compact IP enables the integration of multiple DACs on a single chip while still accommodating SyFaLa and the Faust IP (Table 2).

Table 2. Comparison of this project with State of the Art. The first entry of this project uses an OSR of 128, while the last entry uses an OSR of 1024. With an OSR of 128, the last entry would achieve 147 dB SNR. As the State of the Art was implemented on much older FPGAs, clock frequencies may not be comparable.

Paper	Input WL(bits)	Internal WL(bits)	Slice registers	DSPs	Clock Freq (MHz)	SNR (dB)
Cheung et al.	24	40	3167	-	27.5	140
Ludewig et al.	24	>(24)	566	39	19	138 (sim)
Hongzhi et al.	24	35 /25	918	4	4	115
Ours	16	27/17	397	3	52	122 (sim)[a]
	24	34/17	475	3	52	170 (sim)[a]

[a] This result does not consider the already existing quantisation noise of the Nyquist rate input.

4 Conclusion and Future Work

This paper presented a preliminary result on the implementation and optimisation of an under-microsecond, fifth order, 1 bit SD DAC on FPGA. It is integrated into the SyFaLa toolchain, dedicated to ultra low-latency audio DSP applications. SyFaLa dynamically instantiates the required number of DACs based on the output channels of the provided DSP programme which can be written in Faust or C/C++. Consequently, the DAC is capable of directly playing any audio generated by SyFaLa on-chip within the FPGA. The SD DAC IP implementation, written in VHDL, targets a Digilent Zybo Z7-20 FPGA and requires only a small percentage of the board's resources, leaving a reasonable amount of space for the DSP programme. Simulations indicate that the maximum achievable SNR is approximately 170 dB at 48 kHz SR, using an OSR of 1024, or at an input sample rate of 768 kHz with an OSR of 64. The DAC is configurable in both size and speed, allowing trade-offs between performance and resource usage. For instance, with a 16-bit input, a DAC operating with a 12.288 MHz clock and a 256x OSR achieves an SNR of 101 dB.

We proposed a design-space exploration methodology to search for a trade-off between data path sizes, SDM coefficients and the SNR. However, we assume homogenous WLs, which is conservative. To target a mixed-precision fine-tuned implementation, we plan to provide a mathematical model of the DAC design problem for solving, for example, using integer linear programming, similarly to [9, 22]. The idea is to provide a bit-level model of the computations, including the impact of rounding on each data path, and optimize resource allocation while respecting an output accuracy constraint (not in terms of SNR but ENOBs), following the *computing just right* philosophy.

Computing Just Right. Our approach is to treat the accuracy constraint as a first-class citizen during the design. Indeed, in order to produce reliable circuits, we propose to see the word length specification of the output signal as the specification on the accuracy, guaranteeing faithful rounding. In other words, we wish to guarantee that the computed result behaves as an exactly computed one, rounded just once with an error smaller than the weight of its least significant bit. The cornerstone of such approach is a reliable and tight error propagation model and analysis. This has been solved for the FIR and IIR filters [21] using different types of filter structures [8,20,22] and different approaches to constant multiplication. However, one of the requirements for the target IIR filters is their strict stability. Or, in SD DACs, the oscillators have the poles equal to exactly 1. This major bottleneck would require an additional theoretical study for the error propagation in the overall DAC.

Furthermore, instead of using generic multiplications, we will investigate constant multiplication approaches, e.g., KCM-based [6] or shift-and-add based [7, 9]. These methods optimize complexity, resource usage, and power requirements while further improving latency. For example, [9] mentions a 60% reduction in used LUTs, over 40% reduction in power, and approximately 30% reduction in delay for the shuft-and-add based circuits compared to generic multipliers. Hence, there might be a significant room for improvement for the resulting fixed-point SD DAC implementation.

References

1. Alexandre, L., Lecomte, P., Galland, M.A., Popoff, M.: Feedback acoustic noise control with faust on FPGA: application to noise reduction in headphones. In: Proceedings of the International Faust Conference (SMC/IFC-22)
2. Athar, S., Siddiqi, M.A., Masud, S.: Design and FPGA implementation of a 2nd order adaptive delta sigma modulator with one bit quantization. In: Proceedings of the 2010 International Conference on Field Programmable Logic and Applications, pp. 388–393
3. Aziz, P., Sorensen, H., van der Spiegel, J.: An overview of sigma-delta converters. IEEE Signal Process. Mag. **13**(1), 61–84
4. Chen, K.C., Chang, C.Y., Kuo, S.M.: Active noise control in a duct to cancel broadband noise. In: IOP Conference Series: Materials Science and Engineering 237
5. Cheung, R., Pun, K.P., Yuen, S., Tsoi, K.H., Leong, P.: An FPGA-based re-configurable 24-bit 96khz sigma-delta audio DAC. In: Proceedings of the 2003 IEEE International Conference on Field-Programmable Technology (FPT) (IEEE Cat. No.03EX798), pp. 110–117
6. de Dinechin, F., Kumm, M.: Multiplication by constants. In: Application-Specific Arithmetic: Computing Just Right for the Reconfigurable Computer and the Dark Silicon Era, pp. 365–426
7. Garcia, R., Volkova, A., Kumm, M.: Truncated multiple constant multiplication with minimal number of full adders. In: IEEE International Symposium on Circuits and Systems, ISCAS 2022, Austin, TX, USA, May 27 - June 1, 2022, pp. 263–267. IEEE (2022)

8. Garcia, R., Volkova, A., Kumm, M., Goldsztejn, A., Kühle, J.: Hardware-aware design of multiplierless second-order IIR filters with minimum adders. IEEE Trans. Signal Process. **70**, 1673–1686 (2022)
9. Garcia, R., Volkova, A.: Toward the multiple constant multiplication at minimal hardware cost. IEEE Trans. Circuits Syst. I: Regular Papers, 1–13
10. Hongzhi, S., Jie, W., Yichao, M., Yang, K., Jie, Z.: A FPGA-based 7-order 1-bit sigma-delta modulator for high-precision signal generation. In: Proceedings of the 9th International Conference on Electronic Measurement & Instruments
11. Kong, X., Kuo, S.M.: Study of causality constraint on feedforward active noise control systems. IEEE Trans. Circuits Syst. II: Analog Digital Sig. Process. **46**(2), 183–186
12. Kulka, Z., Woszczek, P.: Implementation of digital sigma-delta modulators for high-resolution audio digital-to-analog converters based on field programmable gate array. Archives of Acoustics **33**(1), 93, number: 1
13. Li, X., Lee, A.: An FPGA implemented 24-bit audio DAC with 1-bit sigma-delta modulator. In: Proceedings of the 2010 IEEE Asia Pacific Conference on Circuits and Systems, pp. 768–771
14. Ludewig, R., et al.: IP generation for an FPGA-based audio DAC sigma-delta converter. In: Proceedings of the 2004 International Conference on Field Programmable Logic and Applications, pp. 526–535
15. Magrath, A., Clark, I., Sandler, M.: Design and implementation of a FPGA sigma-delta power DAC. In: Proceedings of the 1997 IEEE Workshop on Signal Processing Systems. SiPS 97 Design and Implementation formerly VLSI Signal Processing, pp. 511–521
16. Ocampo-Hidalgo, J.J.: System and circuit approaches for the design of multi-mode sigma-delta modulators with application for multi-standard wireless receivers
17. Orlarey, Y., Fober, D., Letz, S.: Faust: an efficient functional approach to dsp programming. New computational paradigms for computer music, pp. 65–96 (2009)
18. Popoff, M., et al.: Audio DSP to FPGA compilation. In: Proceedings of the 2023 IEEE 34th International Conference on Application-specific Systems, Architectures and Processors (ASAP), pp. 31–32
19. Schreier, R., Temes, G.C.: Understanding Delta-Sigma Data Converters
20. Volkova, A., Hilaire, T.: Fixed-point implementation of lattice wave digital filter: Comparison and error analysis. In: 23rd European Signal Processing Conference, EUSIPCO 2015, Nice, France, August 31 - September 4, 2015, pp. 1118–1122. IEEE (2015)
21. Volkova, A., Hilaire, T., Lauter, C.: Arithmetic approaches for rigorous design of reliable fixed-point LTI filters. IEEE Trans. Comput. **69**(4), 489–504
22. Volkova, A., Istoan, M., de Dinechin, F., Hilaire, T.: Towards hardware IIR filters computing just right: Direct form i case study. IEEE Trans. Comput. **68**(4), 597–608
23. Zhang, H., Pandey, A., Wang, D.L.: Low-latency active noise control using attentive recurrent network. IEEE/ACM Trans. Audio Speech Lang. Process. **31**, 1114–1123
24. Zhang, L., Qiu, X.: Causality study on a feedforward active noise control headset with different noise coming directions in free field. Appl. Acoustics **80**, 36–44

AI Accelerators, Architectures and Benchmarking

Automated Design of CNN-Based Image Classifiers for Hailo-8L Accelerator

David Nepras and Lukas Sekanina(✉)

Brno University of Technology, Faculty of Information Technology, Brno, Czech Republic
{xnepra02,sekanina}@fit.vut.cz

Abstract. To provide good classification accuracy and meet challenging design constraints, image classifiers deployed in embedded systems are typically implemented using convolutional neural networks (CNNs) accelerated with specialized inference accelerators. We propose a supernet-based neural architecture search (NAS) method to automate the design process of CNNs executed on the Hailo-8L accelerator, an expanding module for the Raspberry Pi Kit. The method utilizes a multi-objective genetic algorithm in conjunction with a pre-trained supernet and latency predictor to accelerate the evaluation of candidate CNNs. The latency predictor is trained using a collection of generated CNNs whose exact latency is determined directly on the Hailo-8L accelerator. The method is evaluated using CIFAR-100 and ImageNet-100 benchmarks and provides classifiers with competitive accuracy-latency trade-offs.

1 Introduction

This work investigates an efficient deployment of *convolutional neural networks* (CNNs) as image classifiers in embedded systems. To meet challenging design constraints, such systems typically execute pre-trained CNNs on dedicated inference accelerators [19]. Our objective is to design CNNs optimized primarily for accuracy and latency on a specific accelerator, the *Hailo-8L*, an AI module that extends the Raspberry Pi platform. Although CNN-based classifiers for selected benchmark tasks have already been adapted to Hailo-8L [8], and several applications have been reported [1], the availability of a fully automated tool for efficient CNN design would greatly enhance the platform's usability. Such a tool would be especially valuable in scenarios where a user requires (a) a specialized CNN for a new data set to be classified, or (b) multiple optimized CNN implementations (exhibiting different accuracy-latency-energy trade-offs) must be available to select the most suitable one for a given deployment.

The proposed approach is based on hardware-aware *neural architecture search* (NAS) methods [3,12,14,16,18] that allow designers to automate CNN design for given specifications. We utilize a multi-objective genetic algorithm to generate CNN architectures optimized for both accuracy and latency. As evaluating these

M. Kowalczyk and C. Monière (Eds.): DASIP 2026, LNCS 16512, pp. 69–81, 2026.
https://doi.org/10.1007/978-3-032-23871-9_6

metrics is expensive, two techniques are adopted to reduce the design time: (1) weight sharing in a supernet and (2) a latency proxy.

(1) The supernet is a large, over-parameterized CNN that must be trained beforehand [15]. Its sub-networks (subnets) are seen as candidate CNNs for solving the desired problem. They show different trade-offs between accuracy and other objectives. The task is to select the most suitable subnet with respect to the problem specification. The primary advantage of this method is that the subnets share weights with the supernet. When sampled from the supernet, a subnet requires only light fine-tuning (additional training on a smaller data set or a specific data set to fine-tune its accuracy), as opposed to expensive full training.

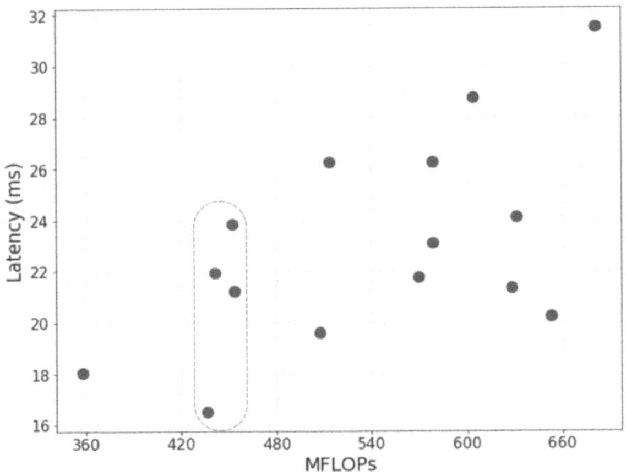

Fig. 1. Latency vs MFLOPS for selected CNNs generated during our experiments and executed on the Hailo-8L for image classification.

(2) Estimating latency as the number of FLOPs, multiply and accumulate (MAC) operations, or trainable parameters (model size), which are usually provided with the reference CNNs implemented for the accelerator, is not reliable [24]. We have measured the properties of several CNNs generated during our preliminary experiments on the Hailo-8L accelerator. Figure 1 shows that CNNs with almost identical MFLOPS can have very different latencies. When Hailo-8L is already programmed to execute a given CNN, its average latency can be quickly obtained by measuring the inference time for several input images and calculating the average. However, in our case, the NAS algorithm will generate many candidate CNNs, for which the latency has to be determined. Obtaining the true inference latency requires 15 min for an average-size CNN due to a slow mapping of the CNN on hardware resources and Hailo-8L reconfiguration. To reduce NAS execution time, we build and train a surrogate model to avoid on-chip latency measurements.

The paper makes the following contributions: (a) As far as we know, this work is the first to introduce a hardware-aware NAS framework specifically tailored for Hailo accelerators. (b) We develop a fast latency prediction model that avoids the need for direct on-device latency measurements. (c) By integrating this predictor with a supernet-based search space and a multi-objective evolutionary optimization strategy, we obtain CNN architectures that achieve competitive accuracy – latency trade-offs on the CIFAR-100 and ImageNet-100 benchmarks.

2 Related Work

NAS: The goal of NAS is to deliver a neural network that maximizes a quality metric (e.g., classification accuracy) for user-provided training and test data sets [14,16] and, optionally, satisfies additional constraints imposed on, for example, model size, latency, or inference energy on a target accelerator. The latter is called a hardware-aware NAS [3,12]. A candidate CNN (only CNNs will be considered in this paper) is usually described as a vector of integers that defines its architecture and parameters. The vector specifies either a complete CNN or a subset of CNN (a block), which is repeatedly applied to build a complete CNN. In *supernet-based methods*, each candidate CNN is seen as a subnet of a larger network, the supernet [15,26]. Once the supernet model is trained, its weights can be reused by any subnet sampled from the supernet. The expensive design of the supernet is amortized by reusing it for different target design scenarios.

The search algorithms (such as evolutionary algorithms, reinforcement learning, or Gaussian process) utilized by NAS methods usually perform a discrete optimization. Examples of evolutionary NAS methods include CNN_GA [23], SI-EvoNet [28], and AE-CNN [22]. However, gradient methods, transforming the inherently discrete design problem into continuous ones, have also been proposed, namely in differentiable NAS methods [13]. To further speed up the NAS method, the accuracy, latency, and inference energy of candidate CNN implementations are often estimated using analytical, simulation, or surrogate model building methods [3,21].

Accelerator: Surveys [3,16] report that CNNs have been optimized with respect to latency, energy, and other hardware metrics, and on various hardware platforms, including GPUs, FPGAs, specialized hardware accelerators, microcontrollers, and mobile phones. The Hailo-8L accelerator is a new platform in this context. It has a powerful and scalable structure-driven data flow architecture in which computing resources (memory, control, and compute blocks) are distributed throughout the chip [7]. For each CNN layer, the Hailo software analyzes the requirements and allocates relevant resources to minimize data movement. The design tools provide proprietary quantization schemes and support common formats such as TensorFlow, PyTorch, and ONNX. To run these models in Hailo-8L, they must be converted to Hailo Executable Format (HEF) using the Hailo Dataflow Compiler. This process involves parsing the model, optimizing it for inference, and quantizing it to an 8-bit integer representation. Since the design software for Hailo accelerators is proprietary, users have no direct control over the mapping of a given CNN onto the accelerator.

3 Proposed Method

Figure 2 gives an overview of the proposed method. To avoid expensive and time-consuming training of a complex supernet, we adopt an existing supernet that is available in the repository [2] associated with AttentiveNAS [26] and AlphaNet [25] NAS methods. The supernet is pre-trained on ImageNet but can be used for other similar image classification data sets.

3.1 Supernet-Based Evolutionary NAS

The supernet from [2] consists, according to its PyTorch model, of 1301 elementary layers such as Conv2d, Linear, ReLU, etc. (18 different layer types reported). Some of them are not configurable, but others exist in different configurations as specified in Table 1. For example, Conv denotes regular convolutional layers, MBConv refers to the inverted residual block [17], MBPool denotes the efficient last stage [9], and SE represents the squeeze and excite layer. The configuration is possible in terms of Width (channel width per layer), Depth (the number of repeated MBConv blocks), Kernel size (3 or 5), and Expansion ratio (ER) for the depth-wise convolution layer in each MBConv block. Four different input resolutions are available. Papers [25,26] employ a single-objective genetic algorithm to generate candidate CNNs. The search is constrained by CNN complexity, which is the maximum FLOPS allowed for one inference. We propose a *multi-objective* genetic algorithm based on NSGA-II, optimizing accuracy and latency together, and providing thus a set of non-dominated solutions (Pareto front) in each run.

Problem Encoding. We follow the encoding proposed in [25,26]. The chromosome specifies a particular configuration for each configurable block, effectively defining the architecture of a subnet. For example, consider MBConv-6 in Table 1. If the corresponding genes of the chromosome are (3, 5, 1), then MBConv-6 is configured: Width = 216; Depth = 8; Kernel size = 5; i.e., the last option is always selected in Table 1 (the first option is encoded 0). Although the total length of the chromosome is only 31 genes, the subnets can be complex because they always contain predefined non-configurable parts.

Fitness Function. As Fig. 2 shows, a candidate CNN denoted α and defined by its chromosome, is evaluated using two fitness functions:

- f_1 = Classification_Accuracy(α) is the accuracy measured on the test set. To do so, α is first fine-tuned for e epochs using a subset of the training data set (see Sect. 5.1).
- f_2 = Latency(α) is estimated using the latency predictor (Sect. 3.2).

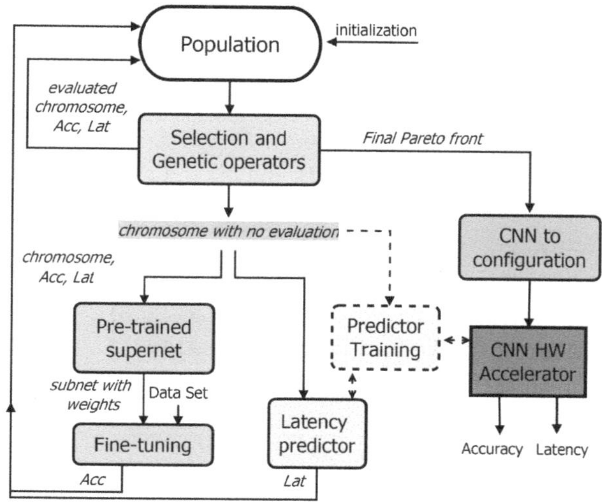

Fig. 2. A supernet-based evolutionary NAS with a latency predictor. Every chromosome (representing a candidate subnet) that has not been evaluated undergoes the evaluation to obtain accuracy (Acc) and latency (Lat). Training of the latency predictor is performed offline (dashed lines). Resulting CNNs are implemented in a hardware accelerator.

Search Algorithm. A multi-objective genetic algorithm NSGA-II, implemented according to [5], optimizes two objectives (f_1, f_2). As NSGA-II evolves a set of non-dominated solutions, it adopts the concept of Pareto dominance, i.e., a solution is said to *dominate* another solution if it is not worse in all objectives and better in at least one. All new offspring networks, created using one-point crossover and mutation, are evaluated (see "chromosome with no evaluation" in Fig. 2). Selected but already evaluated chromosomes are not evaluated again. The algorithm terminates after producing a specified number of generations.

3.2 Latency Predictor

Transforming a CNN model in ONNX format into the internal configuration of the Hailo-8L accelerator, deploying it, running inference, and measuring latency requires approximately 15 min per model evaluation. Consequently, it is impractical to use the accelerator directly within the evaluation loop of NSGA-II. Instead, we employ a latency predictor.

To train the predictor, we construct a data set consisting of subnetworks sampled from the supernet. For each sampled subnet, the ground-truth latency is obtained by executing it on the Hailo-8L accelerator. Due to the high evaluation cost, the resulting data set is relatively small (150 samples). Nevertheless, this data set is sufficient because our method does not require precise latency esti-

Table 1. Configuration options for blocks that constitute the supernet (adopted from [26]). ER – Expansion ratio; SE – Squeeze&Excite layer included.

Block name	Channel width	Depth	Kernel size	ER	SE
Conv	{16, 24}	-	3	-	-
MBConv-1	{16, 24}	{1,2}	{3, 5}	1	N
MBConv-2	{24, 32}	{3, 4, 5}	{3, 5}	{4, 5, 6}	N
MBConv-3	{32, 40}	{3, 4, 5, 6}	{3, 5}	{4, 5, 6}	Y
MBConv-4	{64, 72}	{3, 4, 5, 6}	{3, 5}	{4, 5, 6}	N
MBConv-5	{112, 128}	{3, 4, 5, 6, 7, 8}	{3, 5}	{4, 5, 6}	Y
MBConv-6	{192, 200, 208, 216}	{3, 4, 5, 6, 7, 8}	{3, 5}	6	Y
MBConv-7	{216, 224}	{1, 2}	{3, 5}	6	Y
MBPool	{1792, 1984}	-	1	6	-
Input resolution: {192, 224, 256, 288}					

mation. NSGA-II relies primarily on the relative ranking of candidate solutions rather than on their absolute latency values.

After a preliminary comparison of several regression models, including feed-forward fully connected neural networks (FCNN), Gaussian process regression, and Gradient Boosted Trees, we adopt a simple FCNN for latency prediction. The input layer receives 31 features corresponding to the chromosome encoding of a subnet, followed by two hidden layers with 32 and 16 neurons. The FCNN outputs a single scalar that represents the predicted latency.

4 Setup

Experiments are primarily conducted for the CIFAR-100 data sets, where the goal is to classify 32 × 32 pixel color images into 100 classes [11]. To further evaluate the scalability of the proposed NAS method, we adopted the ImageNet-100 dataset [10] (224 × 224 pixel color images, 100 classes), a subset of the full ImageNet. The default setup for all experiments is as follows.

The *latency predictor* based on the 31-32-16-1 FCNN is trained using PyTorch for up to 2000 epochs, with early stopping based on the validation mean squared error (MSE), using a patience of 50 epochs. The FCNN's performance is periodically evaluated on a test set using both MSE and mean absolute error (MAE) metrics. The training set, test set, and validation set contain feature vectors extracted from 120, 18 and 12 CNNs, respectively. To train the FCNN, we used the Adam optimizer with an initial learning rate of 0.001, a weight decay of 1×10^{-4}, and the Smooth L1 Loss (Huber Loss function). A learning rate scheduler (`ReduceLROnPlateau`) is employed to reduce the learning rate by a factor of 0.5 if the validation loss does not improve for 10 consecutive epochs.

After conducting several test experiments to determine the best parameters of NSGA-II and considering the available computing resources, NSGA-II is set to operate with a population size of 28 individuals evolved for 50 generations, a mutation probability of 0.35, and a crossover probability of 0.9. The candidate subnets sampled by NSGA-II are fine-tuned using PyTorch ($e = 2$) with the Adam optimizer. The learning rate is set to 0.001, the weight decay of 1×10^{-5} is applied for regularization, and the training loss is calculated using the cross-entropy loss function. Fine-tuning is implemented with a batch size of 128.

All experiments are performed on a server (2× Intel Xeon Silver 4314 CPU at 2.40 GHz – 16 cores, 256 GB RAM) equipped with NVIDIA RTX A5000 GPUs. The Hailo-8L neural accelerator is accessed via a Raspberry Pi AI Kit.

5 Results

First, we describe how the NAS algorithm was tuned to reduce the execution time. Next, we present the latency prediction experiments. Finally, we evolve CNNs for CIFAR-100 and ImageNet-100 benchmarks, reporting both accuracy and latency measurements on the Hailo-8L accelerator.

5.1 Optimized Fine-Tuning

As shared weights inherited from the supernet do not often directly lead to CNNs (subnets) showing sufficient accuracy, the subnets have to be fine-tuned for several epochs e. We fine-tune the entire CNN for just one epoch. In the remaining $e - 1$ epochs, only the final fully connected layer is fine-tuned, while the other layers are left untouched. The accuracy is significantly improved after an average of two epochs. This strategy is 4× faster than common fine-tuning. Another speedup is achieved by reducing the training data set to one-third. Compared to the complete training data set, the training period is 1.65× faster and the error drop jointly considered with FLOPs is not statistically significant under a multivariate analysis of variance.

5.2 Latency Prediction

The training and evaluation results confirm that the FCNN-based predictor provides the highest accuracy among the models considered. Specifically, it achieves the lowest mean squared error (MSE = 2.069) on our data set, compared to Gradient Boosted Trees (MSE = 9.543) and Gaussian Process regression (MSE = 27.757). Figure 3 visualizes the correlation between predicted and real latency for CNNs that are considered for training (blue points), testing (green points), and validation (red points) of the FCNN predictor. The Spearman's rank correlation coefficient, which measures the monotonic agreement between predicted and true latency, is 0.908 for the training set and 0.895 for the validation set. These values indicate a strong alignment in the relative ordering of latency values, which is the primary requirement for effective use of the predictor within NSGA-II.

During the NSGA-II runs presented in the following sections, we recorded both the predicted and measured latencies of 119 generated CNNs, none of which were included in the predictor's training set. For this new data set, the Spearman's rank correlation coefficient is 0.881, confirming that the FCNN-based predictor preserves the relative ordering of latencies with high fidelity. Consequently, NSGA-II, when guided by the predicted latency, is able to reliably select low-latency CNN architectures.

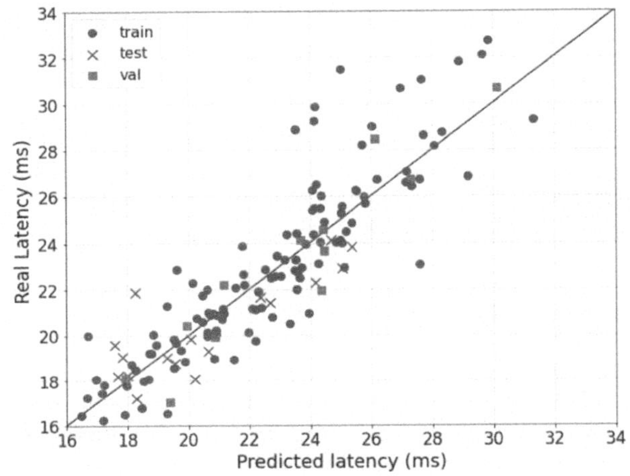

Fig. 3. Correlation between predicted and real latency on Hailo-8L for examined CNNs on training, test, and validation data sets.

5.3 Evolved CNNs

CIFAR-100: From the NSGA-II runs on CIFAR-100, we select four candidate CNNs (the one with the highest accuracy, the one with the lowest latency, and two that represent strong accuracy – latency trade-offs), and then perform an additional 5 training epochs on the full training set. As shown in Table 2, the accuracy of these selected models (denoted SCF100-1 to SCF100-4) improves substantially, by approximately 12 – 14%. We report only accuracy as the quality indicator, following standard practice in the relevant literature.

We compare the evolved CNNs with several widely used, human-designed architectures (VGG16 [20], MobileNetV2x1_4 [17], and RepVGG [6]). Their PyTorch implementations were obtained from the repository [4], compiled, and executed on the Hailo-8L accelerator for evaluation (Table 2). The evolved CNNs achieve higher accuracy than these human-designed baselines. However, when maintaining competitive accuracy, their latency is generally higher, which can be attributed to constraints inherited from the supernet used during search. As illustrated in Fig. 4, the evolved CNNs introduce new design points in the accuracy – latency trade-off space, demonstrating distinct and valuable alternatives to existing architectures.

A direct comparison with related evolutionary NAS methods is not feasible, as these approaches do not target hardware accelerators, and the resulting CNN models are not publicly available. Considering only accuracy, parameter count, and execution time (Table 2), SI-EvoNet [28] outperforms other NAS methods. Compared to AE-CNN [22], our method produces models with more favorable performance characteristics and does so more efficiently. Xie et al. [27] and CNN_GA [23] generate smaller networks than ours, and the approach by Xie et al. [27] also achieves faster execution. Overall, our method yields CNNs with good accuracy, albeit with larger model sizes, and with moderate execution times. Importantly, the used multi-objective algorithm allows us to obtain, in a single run, a diverse set of architectures representing different accuracy – latency trade-offs, whereas other considered NAS methods typically report only a single best model.

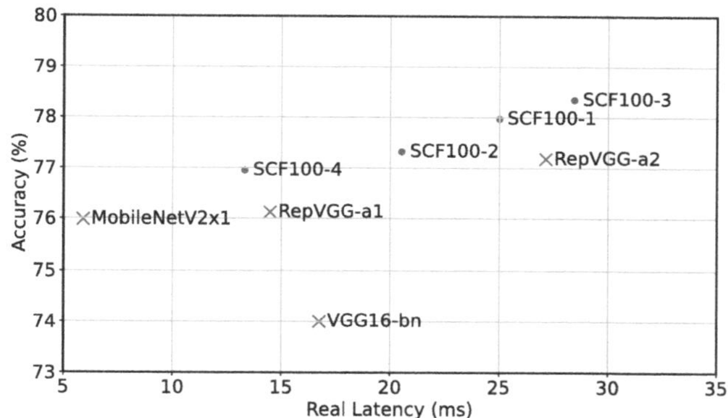

Fig. 4. Comparison of evolved (blue dots) and human-created (green crosses) CNNs accelerated on Hailo-8L and examined on the CIFAR-100 benchmark. (Color figure online)

ImageNet-100: Due to the greater complexity of this data set, we train candidate CNNs on the full training set for two epochs to get their accuracy fitness rather than relying on fine-tuning. After obtaining the final Pareto front from NSGA-II, we select three distinct CNNs for detailed evaluation. These CNNs are fine-tuned for an additional four epochs, and their true latency is measured on the accelerator.

Table 3 compares the selected CNNs with several resource-optimized variants of MobileNetV2 that were obtained by reducing the depth of the network and modifying the bottleneck components [29]. Since their results on ImageNet-100 are provided for a standard GPU, we can compare these models with our evolved

Table 2. Comparison of evolved subnets (SCF100-1 – SCF100-4) with selected conventional CNNs from [4] and evolutionary NAS methods (on CIFAR-100 benchmark). Subnets' accuracy is given at the end of NAS (Acc-NAS) and after additional retraining (Acc). Pred-Lat and Real-Lat denote the predicted latency and real latency on Hailo-8L.

CNN/ NAS	Acc-NAS (%)	Acc (%)	Pred-Lat [ms]	Real-Lat [ms]	Perf. [MFLOPs]	Params. $\times 10^6$	t_{GPU} [day]
SCF100-1	65.61	77.97	23.78	25.00	468.22	11.91	3.6
SCF100-2	63.29	77.32	20.62	20.52	327.62	8.87	3.6
SCF100-3	66.26	78.34	28.53	28.45	519.21	17.88	3.6
SCF100-4	59.50	76.95	13.87	13.32	342.06	9.16	3.6
VGG16-bn [4]	–	74.00	–	16.74	313.77	15.29	–
MobNetV2x1_4 [4]	–	75.98	–	5.90	170.23	4.49	–
RepVGG-a1 [4]	–	76.12	–	14.48	851.44	12.93	–
RepVGG-a2 [4]	–	77.18	–	27.15	1850.22	26.94	–
CNN_GA [23]	–	77.97	–	NA	NA	4.1	40
SI-EvoNet [28]	–	79.16	–	NA	NA	0.99	0.81
Xie et al. [27]	–	77.02	–	NA	NA	3.4	0.92
AE-CNN [22]	–	77.98	–	NA	NA	20.9	10

CNNs (SIN100-1 to SIN100-3) in terms of accuracy, performance, and model size only. The evolved CNNs are approximately 3.9× larger, but they also achieve substantially higher accuracy (an average improvement of +4.9%), primarily because the supernet allows for sampling more expressive subnet architectures. However, the models from [29] are not publicly available, so a direct latency comparison is not possible.

Table 3. Comparison of evolved subnets (SIN100-1 – SIN100-3) with three versions of MobileNetV2 [29] on ImageNet-100.

CNN	Acc-NAS	Accuracy	Pred-Lat	Real-Lat	Perf.	Params.
SIN100-1	79.93	80.97	14.14	17.49	401.76	10.49
SIN100-2	82.40	83.18	25.05	27.78	850.60	12.85
SIN100-3	84.55	85.12	28.19	30.82	908.43	13.98
MobileNetV2(6)	–	77.86	–	NA	314.19	3.50
MobileNetV2(5)	–	78.52	–	NA	298.51	3.18
MobileNetV2(4)	–	78.21	–	NA	282.83	2.86

6 Conclusions

We proposed and evaluated a hardware-aware NAS method utilizing a latency predictor for the automated design of CNN-based image classifiers targeting the Hailo-8L accelerator. After tuning the algorithm, the evaluation was performed on CIFAR-100 and ImageNet-100 data sets.

For CIFAR-100, the proposed evolutionary NAS effectively exploits the available resources of the supernet and delivers network architectures showing high accuracy after only moderate retraining. In our comparisons, the human-created MobileNetV2 shows the smallest latency because it is designed for mobile applications; however, its classification accuracy is far behind the CNNs evolved for this benchmark. Overall, our method yields CNNs with good accuracy, albeit with larger model sizes, and with moderate execution times compared to relevant evolutionary NAS methods.

In the case of ImageNet-100 and when compared to MobileNetV2, the accuracy of evolved CNNs is higher. However, we cannot compare latency because relevant MobileNetV2 models trained for ImageNet-100 are not available.

As the proposed method does not assume anything about the accelerator's architecture, it can be used with other accelerators after building a surrogate model for latency and other metrics if the objective is to include them in the optimization. However, to reach competitive CNNs, it is crucial to choose a suitable supernet (and a corresponding sampling strategy) with respect to the target classification problems and the accelerator. The supernet used in this work tends to generate unnecessary complex models that show competitive accuracy.

Our future work will primarily be devoted to evaluating the proposed method on data sets from real-world applications and improving the proposed evolutionary NAS algorithm to reach more efficient implementations of CNNs on the target accelerator.

Acknowledgements. This work was supported by the Czech Science Foundation project 24-10990S.

References

1. Achmadiah, M.N., Ahamad, A., Sun, C.C., Kuo, W.K.: Energy-efficient fast object detection on edge devices for IoT systems. IEEE Internet Things J. 1–15 (2025)
2. AlphaNet: AlphaNet: Improved training of supernet with Alpha-divergence (2025). https://github.com/facebookresearch/AlphaNet
3. Benmeziane, H., El Maghraoui, K., Ouarnoughi, H., Niar, S., Wistuba, M., Wang, N.: Hardware-aware neural architecture search: survey and taxonomy. In: Proceeding of the 30 International Joint Conference on Artificial Intelligence, IJCAI-21, pp. 4322–4329 (2021)
4. Chenyaofo: PyTorch CIFAR models (2025). https://github.com/chenyaofo/pytorch-cifar-models?tab=readme-ov-file
5. Deb, K., Pratap, A., Agarwal, S., Meyarivan, T.: A fast and elitist multiobjective genetic algorithm: NSGA-II. IEEE Trans. Evol. Comput. **6**(2), 182–197 (2002)

6. Ding, X., Zhang, X., Ma, N., Han, J., Ding, G., Sun, J.: RepVGG: making VGG-style convnets great again. In: 2021 IEEE/CVF Conference on Computer Vision and Pattern Recognition (CVPR), pp. 13728–13737 (2021)
7. Hailo: Hailo AI accelerators (2025). https://hailo.ai/products/ai-accelerators/
8. Hailo: Hailo Model Zoo (2025). https://github.com/hailo-ai/
9. Howard, A., et al.: Searching for MobileNetV3. In: 2019 IEEE/CVF International Conference on Computer Vision (ICCV), pp. 1314–1324 (2019). https://doi.org/10.1109/ICCV.2019.00140
10. ImageNet100: Kaggle imagenet-100 dataset (2025). https://www.kaggle.com/datasets/ambityga/imagenet100
11. Krizhevsky, A.: Learning multiple layers of features from tiny images. Technical Report (2009)
12. Lin, Y., Yang, M., Han, S.: NAAS: Neural accelerator architecture search. In: 58th ACM/ESDA/IEEE Design Automation Conference (2021)
13. Liu, H., Simonyan, K., Yang, Y.: DARTS: differentiable architecture search. In: 7th International Conference on Learning Representations, ICLR 2019, New Orleans, LA, USA, May 6-9, 2019. OpenReview.net (2019)
14. Liu, Y., Sun, Y., Xue, B., Zhang, M., Yen, G.G., Tan, K.C.: A survey on evolutionary neural architecture search. IEEE Trans. Neural Networks Learn. Syst. **34**(2), 550–570 (2023). https://doi.org/10.1109/TNNLS.2021.3100554
15. Pham, H., Guan, M.Y., Zoph, B., Le, Q.V., Dean, J.: Efficient neural architecture search via parameter sharing. In: Proceeding of the 35th International Conference on Machine Learning, ICML 2018. vol. 80, pp. 4092–4101. PMLR (2018)
16. Ren, P., et al.: A comprehensive survey of neural architecture search: Challenges and solutions. ACM Comput. Surv. **54**(4) (2021). https://doi.org/10.1145/3447582
17. Sandler, M., Howard, A., Zhu, M., Zhmoginov, A., Chen, L.C.: MobileNetV2: Inverted residuals and linear bottlenecks. In: 2018 IEEE/CVF Conference on Computer Vision and Pattern Recognition, pp. 4510–4520 (2018). https://doi.org/10.1109/CVPR.2018.00474
18. Sekanina, L.: Neural architecture search and hardware accelerator co-search: a survey. IEEE Access **9**, 151337–151362 (2021)
19. Silvano, C., et al.: A survey on deep learning hardware accelerators for heterogeneous hpc platforms. ACM Comput. Surv. **57**(11) (2025). https://doi.org/10.1145/3729215
20. Simonyan, K., Zisserman, A.: Very deep convolutional networks for large-scale image recognition. In: 3rd International Conference on Learning Representations (2015)
21. Song, X., et al.: Efficient evaluation methods for neural architecture search: a survey. IEEE Trans. Artif. Intell. **5**(12), 5990–6011 (2024). https://doi.org/10.1109/TAI.2024.3477457
22. Sun, Y., Wang, H., Xue, B., Jin, Y., Yen, G.G., Zhang, M.: Surrogate-assisted evolutionary deep learning using an end-to-end random forest-based performance predictor. IEEE Trans. Evol. Comput. **24**(2), 350–364 (2020)
23. Sun, Y., Xue, B., Zhang, M., Yen, G.G., Lv, J.: Automatically designing CNN architectures using the genetic algorithm for image classification. IEEE Trans. Cybern. **50**(9), 3840–3854 (2020)
24. Sze, V., Chen, Y.H., Yang, T.J., Emer, J.S.: How to evaluate deep neural network processors: TOPS/W (alone) considered harmful. IEEE Solid-State Circuits Mag. **12**(3), 28–41 (2020). https://doi.org/10.1109/MSSC.2020.3002140

25. Wang, D., Gong, C., Li, M., Liu, Q., Chandra, V.: Alphanet: improved training of supernets with alpha-divergence. In: Proceedings of the 38th International Conference on Machine Learning. vol. 139, pp. 10760–10771. PMLR (2021)
26. Wang, D., Li, M., Gong, C., Chandra, V.: Attentivenas: improving neural architecture search via attentive sampling. In: 2021 IEEE/CVF Conference on Computer Vision and Pattern Recognition (CVPR), pp. 6414–6423 (2021)
27. Xie, Y., Chen, H., Ma, Y., Xu, Y.: Automated design of CNN architecture based on efficient evolutionary search. Neurocomputing **491**, 160–171 (2022)
28. Zhang, H., Jin, Y., Cheng, R., Hao, K.: Efficient evolutionary search of attention convolutional networks via sampled training and node inheritance. IEEE Trans. Evol. Comput. **25**(2), 371–385 (2021)
29. Zhao, L., Wang, L., Jia, Y., Cui, Y.: A lightweight deep neural network with higher accuracy. PLoS ONE **17**(8), 1–15 (2022)

Benchmarking Deep Learning Convolutions on Energy-Constrained CPUs

Enrique Galvez(✉), Adrien Cassagne, Alix Munier, and Manuel Bouyer

Sorbonne Université, CNRS, LIP6, 75005 Paris, France
enrique.galvez@lip6.fr

Abstract. This work evaluates State-of-the-Art convolution algorithms for CPU-based CNN inference. Although most prior studies focus on GPUs or NPUs, CPU implementations remain comparatively under-optimized. Our first contribution is to provide fair benchmarking for embedded CPU inference. We evaluate direct, GEMM-based, and Winograd convolutions across modern CPUs from ARM®, Intel®, AMD®, and NVIDIA® vendors, considering both latency and energy efficiency. To the best of our knowledge, this is the first study to present a fair, cross-vendor comparison of CPU energy consumption using a high-resolution socket-level measurement platform. To validate our methodology, we further compare socket-level power measurements with estimates derived from model-specific registers (MSRs), finding that MSRs underestimate the power consumption of convolution inference by 10–30%. Our results show that the ARM® Cortex-A78AE CPU combined with an implicit GEMM convolution implementation offers the best trade-off between latency and power consumption, achieving ResNet50v1.5 inference in 102 ms with an average power of 25.3 W, corresponding to 2.58 J.

Keywords: Convolution algorithms · Benchmarking · Edge AI · Energy-constrained CPUs · Energy measurements · Compute intensive

1 Introduction and Related Work

Deep neural networks (DNNs) have become pervasive in modern embedded computer vision systems. Among them, convolutional neural networks (CNNs) remain the backbone of most classification [14] and detection pipelines [6], thanks to their relatively low computational and memory requirements as well as the maturity of available software frameworks [12,28]. Even with the rise of Transformer-based models such as DETR [2], CNN backbones still account for a significant portion of inference time: up to 30% in some configurations [18]. As a consequence, convolutions remain a central performance bottleneck in embedded inference of deep learning models.

The convolution operator is the key enabler of automatic feature extraction, but it also dominates the computational workload and energy demand during

M. Kowalczyk and C. Monière (Eds.): DASIP 2026, LNCS 16512, pp. 82–94, 2026.
https://doi.org/10.1007/978-3-032-23871-9_7

both training and inference [22]. In this paper, we show that convolutions account for more than 90% of the inference of popular networks such as ResNet50v1.5, VGG19 and GoogLeNet (see Sect. 4). This cost has motivated extensive research into optimized implementations for GPUs [17,23,25] and CPUs [5,7,21]. However, most of these studies target high-performance hardware or specific architectures, and little attention has been given to embedded CPUs, which are still widely deployed in systems where accelerators are absent or offloading is impractical due to small models or input sizes.

In this article, we address this gap by systematically benchmarking the inference of CNNs on CPUs. We consider CPUs commonly used in embedded and battery-powered systems from the main vendors: ARM®, Intel®, AMD® and NVIDIA®. Our study focuses on three complementary performance metrics: latency, power and energy consumption. These metrics are particularly critical in embedded contexts where energy efficiency and thermal constraints directly affect system design. This study extends the work of Dolz et al. [4], which was limited to ARM-based CPUs. Furthermore, we introduce a protocol leveraging a novel high-resolution power measurement system [3], enabling fine-grained characterization of the full system's energy profile during convolution execution. This work includes the following key contributions: (i) The characterization of a novel protocol based on high-frequency socket power measurement for benchmarking convolutions and the comparison of socket-measured power with power estimated using model-specific registers (MSRs); (ii) A fair and cross-vendor evaluation of State-of-the-Art convolution algorithms on modern embedded CPUs, supported by a multidimensional benchmark considering both inference latency and power consumption in order to guide algorithmic and hardware choices.

The remainder of this article is organized as follows. Section 2 presents the convolution algorithms evaluated in this study. Section 3 describes the hardware and software platforms considered. Section 4 outlines the experimental protocol. Section 5 compares socket-based power measurements with estimates obtained from MSRs. Section 6 analyzes the energy consumption of the convolution algorithms across the studied CPUs. Section 7 provides a benchmark of CNN inference across these CPUs. Finally, Sect. 8 concludes the article.

2 Convolution Algorithms

The convolution operation consists of applying a filter with learnable values across an image. We consider 2-dimensional convolutions over 4-dimensional tensors, which are used by most CNNs in the context of computer vision. Moreover, since convolutions in such networks are rarely dilated or strided, we limit our study to undilated and unstrided convolutions.

Using the notation described in Table 1, the dimensions of the input tensors can be written as: $MB \times IC \times IH \times IW$ and the dimensions of the weight tensors can be written as: $OC \times IC \times KH \times KW$. Following the same notations, the convolution operation can be described formally as:

Table 1. Notation for the main convolution parameters.

MB	batch size	IH, IW	input height, width
KH, KW	kernel height, width	OH, OW	output height, width
IC, OC	input, output channels	PH, PW	padding height, width

$$dst[mb, oc, oh, ow] = bias[oc] + \sum_{ic=0}^{IC-1} \sum_{kh=0}^{KH-1} \sum_{kw=0}^{KW-1} src(mb, ic, ih, iw) \cdot weights[oc, ic, kh, kw], \quad (1)$$

where $ih := oh + kh - PH$ and $iw := ow + kw - PW$.

The output tensor *dst* is computed given an input tensor *src* and using learned values for *bias* and *weights* tensors. Each element of the output tensor is computed as a weighted sum of elements from the input tensor across the three dimensions of the kernel: channels (IC), kernel height (KH) and kernel width (KW). Figure 1 describes the convolution operator.

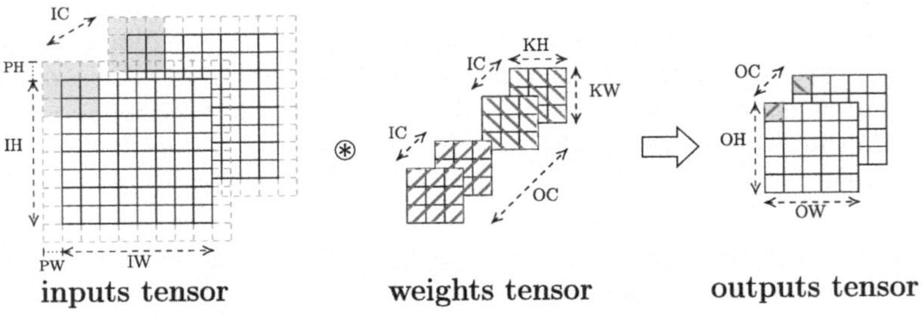

Fig. 1. Convolution Product.

In this article, we aim to minimize the latency of one CNN inference so we consider the batch size to be equal to 1. Predominant CNNs have convolution layers with 3×3 kernels ($KH = KW = 3$), 1×1 kernels and, less commonly, bigger kernels such that 7×7, for example. Later, we detail in Sect. 4 the temporal impact of different convolutions on the inference of common CNNs. The 1×1 convolutions can be implemented directly as a General Matrix Multiplication (GEMM), which is often preferred for performance. Other convolution kernels require specific implementations, discussed in the following paragraphs. Our study proposes a benchmark of State-of-the-Art's three most commonly used implementations of convolutions in modern deployment of CNN inference: direct, GEMM-based and winograd, described as follows.

Direct Method [`direct`]. The convolution operator can be naively implemented using nested loops iterating through the output pixels and corresponding kernels. However, by reordering the loops and grouping the indices iterating over $[0, OC]$, $[0, IC]$, and $[0, OW]$ into fixed-size blocks, Zhang et al. [26] developed an optimized implementation of this computation, which we refer to as `direct`.

Explicit and Implicit Lowering [`im2row` and `gemm`]. These methods convert a convolution operation into a General Matrix Multiplication (GEMM) using "lowering" techniques. The principle is to replicate data in order to transform input and weight tensors into matrices whose product corresponds to the convolution output. This approach can be particularly efficient, as GEMM is well optimized on modern CPUs [13] and GPU-based accelerators [1]. The lowering procedure is presented in Fig. 2 and we refer to our explicit lowering convolution as `im2row` in the following. Although `im2row` involves a large overhead due to data movements, implicit lowering reduces this overhead by transforming small tiles of the tensor on-the-fly and computing the output result for each tile in parallel [17]. The default implementation of OneDNN [9] uses implicit lowering and is referred to as `gemm` in the following. These methods are particularly efficient on highly parallel architectures such as GPUs or TPUs [27].

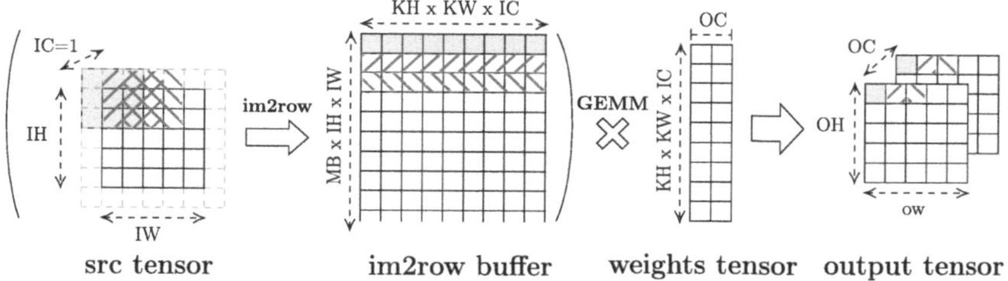

Fig. 2. Computing a convolution using `im2row`.

Winograd Convolution [`wino`]. The last method is an extension of Winograd's algorithm [24] which reduces the number of floating-point multiplications of one-dimensional convolutions by taking advantage of redundancy patterns, at the cost of increasing the number of additions. To transpose this method to deep-learning convolutions, we followed the approach of Lavin and Gray [15], working on 3×3 convolution kernels. This algorithm is referred to as `wino` and its main benefit is to reduce the number of critical floating-point multiplications by 2.25 when the window size is 2. Such methods were implemented on GPUs [20], but are known to show floating-point imprecision with quantized data [19].

3 Experiments Platform

Our study targets CPUs typically deployed in embedded, battery-powered systems from the major processors vendors (ARM®, Intel®, AMD®, NVIDIA®). We selected the processors described in Table 2 for executing our workloads. The first class of CPUs considered in this study comprises the processors integrated into NVIDIA Jetson single-board computers, which are designed for computer vision. The CPUs of the AGX Orin and AGX Xavier boards both implement the ARM v8.2 instruction set: AGX Orin has a ARM Cortex-A78AE CPU and AGX Xavier has a NVIDIA Carmel CPU. A second class includes x86 processors. We evaluate a Ryzen 7 7840U based on the Zen 4 architecture and a Ryzen AI 9 HX 370 based on Zen 5. Finally, we benchmark an Intel Core Ultra 9 185H CPU from the Meteor Lake generation. Some of the studied architectures have heterogeneous CPUs, with performance cores, efficiency cores or low power efficiency cores (respectively p, e and LPe cores). p-cores are the fastest CPU cores but they are more energy consuming than e and LPe cores that are slower. Some CPUs also allow 2-way simultaneous multithreading (2-SMT) on some of their cores. All platforms use Ubuntu distribution. The corresponding Linux kernels are reported in Table 2.

Table 2. Description of the targets with $\mathcal{T}$ the core type and $\mathcal{C}$ the cores number.

				CPU core					Software version	
Product & *Tag*	Vendor	TDP (Watts)	Release	$\mathcal{T}$	Architecture	$\mathcal{C}$	SMT	Proc. (nm)	Linux (Ubuntu)	GCC
Jetson AGX Xavier RAM 16 GB *(NV Carmel)*	NVIDIA	30 ≤	Oct'18	p	Carmel	8	1	12	4.09.201	11.4.0
Jetson AGX Orin RAM 64 GB *(ARM Cortex)*	ARM	60 ≤	Mar'23	p	Cortex-A78AE	12	1	8	5.10.120	11.4.0
Ryzen 7 7840U RAM 32 GB *(AMD Zen4)*	AMD	15–30	Jan'24	p	Zen 4	8	2	5	6.08.000	13.3.0
Core Ultra 9 185H RAM 32 GB *(Intel U9)*	Intel	35–115	Jul'24	LPe	Crestmont	2	1	5	6.14.000	14.2.0
				e	Crestmont	8	1	7		
				p	Redwood Cove	6	2	7		
Ryzen AI 9 HX 370 32 GB *(AMD Zen5)*	AMD	15–54	Oct'24	e	Zen 5c	8	2	4	6.08.000	13.3.0
				p	Zen 5	4	2	4		

The algorithms described in Sect. 2 are implemented[1] into Intel's OneDNN framework version 3.4 [9]. This framework provides an abstraction level that allows one to implement deep learning primitives, with a built-in correctness and performance evaluation tool called BenchDNN. We compiled OneDNN with the `-fopenmp -O3` optimization flags. OpenMP runs threads to leverage tensor parallelism. The measurements have been first conducted using BenchDNN, focusing only on the convolutions. Later, a full inference across the ResNet50v1.5 network [8] is evaluated. Its execution relies on ONNX Runtime that links with our OneDNN implementations. ONNX Runtime v1.22.2 was used for all architectures except for NV Carmel which used version 1.17.0 for compatibility. It is worth mentioning that inter-operation parallelism is disabled. Only tensor parallelism inside OneDNN convolutions remains.

[1] The code of our implementations is available at https://github.com/EnriqueGlv/Benchmarking-DL-Convolutions-on-CPUs.

4 Experiments Protocol

The aim of this study is to benchmark latency and energy consumption of CPUs from several vendors for the inference of CNNs. An important thing to note in that we consider both convolution-level benchmark using BenchDNN and network-level inference benchmark using ONNX Runtime. In this context, Fig. 3 compares the temporal cost of several convolutions over the inference of three popular CNN networks: ResNet50v1.5, VGG19 and GoogLeNet. The results have been obtained using ONNX Runtime built-in profiler on the AMD Ryzen7 7840U CPU. In order to measure per-layer timings, we set ONNX Runtime's `GraphOptimizationLevel` to `ORT_DISABLE_ALL` to disable layer fusion.

A first observation we can make from Fig. 3 is that convolutions clearly dominate inference time of evaluated CNNs: often more than 90% of inference time. Another observation is that 1×1 convolutions cost less than 3×3 convolutions individually but the high number of 1×1 convolutions in ResNet makes their cost non-negligible (42% of inference time). Despite the high cost of 1×1 convolutions in ResNet50v1.5, 3×3 convolutions cost approximately 52% of inference time. Another interesting observation is that predominant 3×3 convolutions have similar kernels between the networks. In particular, the most expensive convolutions of ResNet50v1.5 and VGG19 share the same dimensions. As a consequence to these observations, our convolution-level benchmarking will only focus on the most expensive 3×3 convolution in terms of computations: `MB1_IC256IH14_OC256OH14_KH3PH1` (which we refer to as `3x3_IC256_OC256` in Fig. 3 and in the following). In the previous description format, each convolution parameter (from Table 1) is followed by its value. For instance, `MB1` means that minibatch is 1 corresponding to inference without batching.

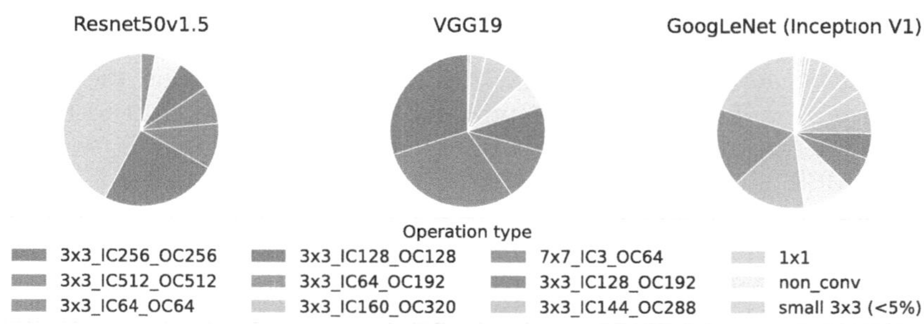

Fig. 3. Relative latency of convolutions on CNN inference depending on their dimensions running on AMD Ryzen7 7840U CPU (AMD Zen4).

One of the contributions of this work is the experimental methodology. While its steps are not new on their own, the novelty of our approach resides in the use, for the first time, of high frequency measurement boards for monitoring socket power consumption of convolutions. The key methodology steps are:

1. Using `taskset` system command, we run the main program on a specific set of CPU HW threads (compact threads pinning on cores or HW threads).
2. We run a warm-up round of 200 convolutions or inferences, in order to ensure the stability of execution speed (stabilization of the CPU frequency).
3. We create a background thread that collects the energy measurements. It is pinning on a unused core when possible.
4. We run 1000 iterations of convolutions or inferences and gather the speed metrics returned by BenchDNN or ONNX Runtime
5. We stop the background thread and compute statistics over the energy measurements.

We used this methodology for all the following measurements, to ensure their pertinence and reproducibility. Section 5 analyzes the pertinence of socket measurements and provide important insights over the measurements uncertainty that can be expected from experimental results.

5 MSRs Versus Socket Measure

To motivate our choice of high-frequency socket measurements over MSRs for power evaluation, we begin by comparing both methods. A first difference resides in the way that power is estimated in both cases. In one hand, MSR power is estimated by reading a register containing the energy consumption of the exposed power domain. By reading these registers at a certain frequency, the average power consumption is estimated for each interval. On the other hand, our high-frequency measurement system measures energy consumption between the power supply and the socket at 1000 samples/s for x86 CPUs [3] or 5000 samples/s for Jetson boards.

In order to evaluate the differences between MSRs power estimates and socket power measurements in the context of convolutions, we performed rounds of experiments for the `3x3_IC256_OC256` convolution, following the protocol described in Sect. 4. Since NVIDIA provides `jetson-power-tools` to precisely report power consumption across the components of Jetson boards, we used it to gather MSR values at 10 samples/s. For x86 CPUs, MSR power is gathered using `perf stat` to read the RAPL `/power/energy-pkg` event at 1 sample/s. We chose these sampling rates to minimize errors due to measurement overhead and the lack of energy updates [11].

Figure 4 provides a comparison between MSRs-based and socket-based power measurements for the inference of the `3x3_IC256_OC256` convolution for ARM Cortex A78AE (Jetson Orin) and AMD Ryzen AI 9 HX370 (x86) CPUs. A first observation we can make is that in both cases, measured MSR power is lower than socket power. However, the difference between socket power and Jetson power does not depend on multithreading while the difference between RAPL and socket power increases with multithreading. A possible explanation for RAPL's behavior is the fact that RAM is not part of the exposed power domain. As RAM usage increases with parallelism, this may explain the increasing difference

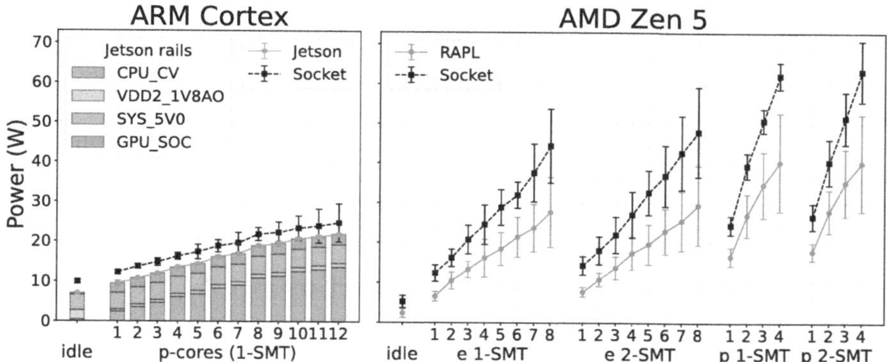

Fig. 4. MSRs versus socket measure of average power and dispersion (standard deviation) depending on multithreading.

between RAPL and socket measurements for AMD Zen5. In addition to RAM, RAPL does not capture power supply losses, fans, storage or network, which may result in a non-constant overhead [10]. A last observation is that standard deviation of power measurements is high for both MSRs and socket readings.

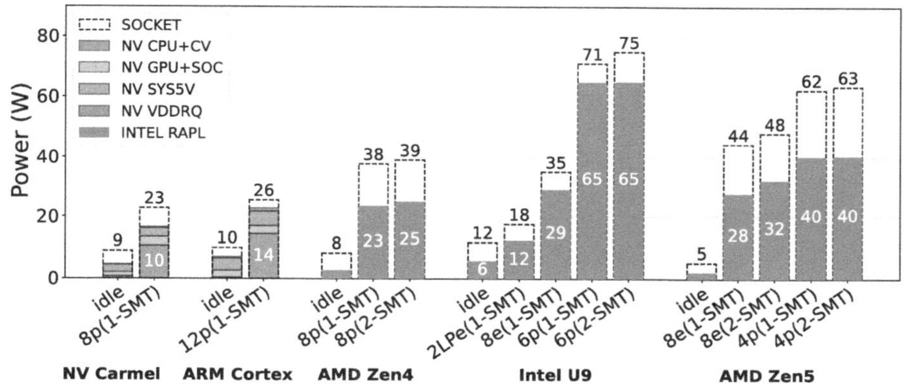

Fig. 5. Energy consumption measurements: MSRs and socket.

Using the same protocol, we extended the results displayed in Fig. 4 to all studied architectures in Fig. 5 the difference between MSRs and socket power measurements. Based on these experiments, we conclude that MSRs-based power measurements do not fully capture the total power consumption measured at the socket. In idle state, MSRs readings are more than 50% lower than the corresponding socket measurements, while during convolution computations, this discrepancy ranges between 10% and 30%. Another observation is that 2-SMT have a small and positive impact over average power consumption.

One may notice that `jetson-power-tools` provides the closest approximation of the actual power consumption (compared to RAPL). Despite the high dispersion of power measurements for both MSRs and socket-based readings, the mean power is more consistent across repeated socket measurements than across repeated MSRs measurements. Consequently, for the following experiments of this article, we rely exclusively on socket measurements to accurately benchmark the power consumption of each system.

6 Convolution Performance Depending on Algorithm

Figure 6 presents the energy consumed to compute the `3x3_IC256_OC256` convolution depending on the number of CPU cores. Additionally, Fig. 6 also compares the different algorithms. An important remark is that uncertainty propagated to energy is way smaller than uncertainty over power, due to the low variance of latency measurements.

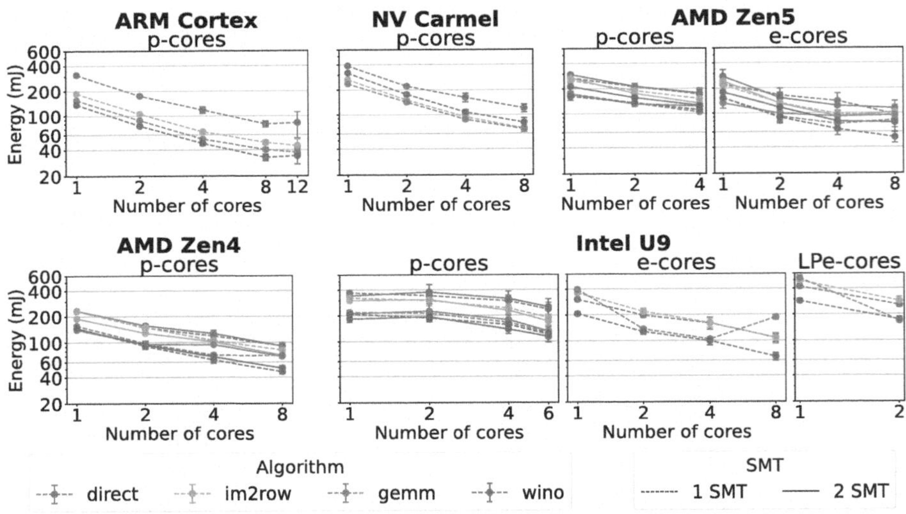

Fig. 6. Energy consumption of `3x3_IC256_OC256` convolution depending on algorithm and multithreading.

First, we observe that increasing the number of physical cores almost always decrease the energy consumption of the convolution. The reason is that increasing the number of cores further decreases latency than it increases power consumption. We also observe that 2-SMT has no significant impact on the average energy consumption of an operation. A possible explanation is that convolution kernels already exert high pressure on backend execution resources and the memory subsystem. However, SMT is most beneficial when execution resources are underutilized, and it may have limited opportunity to improve throughput under such conditions, resulting in negligible energy differences. A surprising result is that e-cores and LPe-cores consumes more than p-cores for the heterogeneous architectures. The reason is that low-power processors are much slower than they are energy efficient. Regarding the algorithms, `wino` and `gemm` are the most energy efficient because they are the fastest on almost every architecture. The main result of this benchmark is that the best algorithm-architecture configurations use 8 p-cores of the ARM Cortex A78AE or the AMD Zen 4, with the `wino` algorithm. Both configurations reach a power consumption that is less than 60 mJ.

7 Cross-Vendor CPU Benchmark for CNN Inference

Results shown in Fig. 7 are measured through a full inference of ResNet50v1.5. Since we target energy constrained devices, power consumption may be a discriminant criteria for the architectures, as well as inference latency. All the inference runs are executed on the same 640 × 640 image from COCO dataset [16] and we ensured the correctness of the predictions.

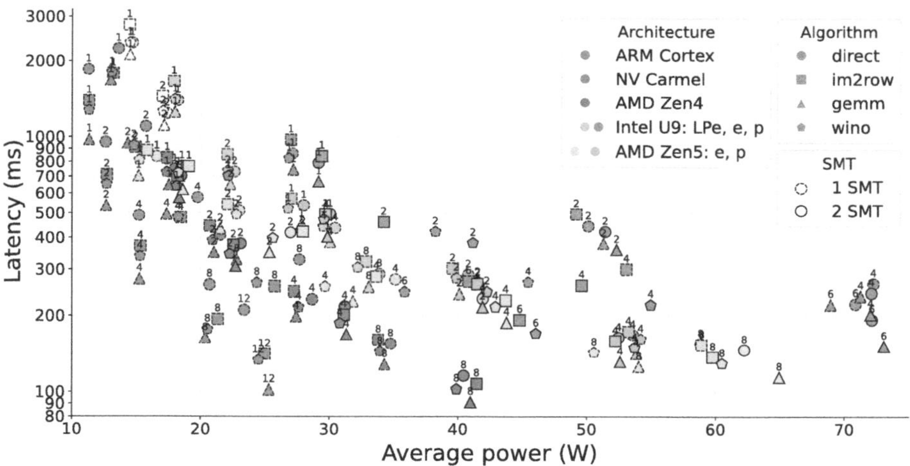

Fig. 7. Latency and power consumption depending on architecture, multithreading and algorithm for a full inference of ResNet50v1.5.

A first observation is that focusing on the computational part of the convolution allows `wino` to take advantage of its arithmetic complexity optimization (see Fig. 6) while in a full inference, significant data management acts in favour of implicit GEMM (`gemm`) implementation. In addition to that, Fig. 7 clearly shows the existence of a trade-off between latency and average power consumption in our context. Moreover, among the studied architecture, ARM Cortex

A78AE is the architecture offering the best trade-off between inference latency and average power consumption. If power budget is higher, AMD Zen4 can allow a faster inference. Additionally, NVIDIA Carmel CPU allow slow inference with a low power budget while, on the contrary, AMD Zen5 offers good latency at the cost of high power consumption. However, AtomMan X7 Ti offers sub-optimal performances on each one of its CPU coretypes.

8 Conclusion

This work introduces an original methodology based on high-resolution socket-level power measurements to benchmark CNN inference on CPUs commonly used in embedded and battery-powered systems. By extending prior studies to the major CPU vendors ARM®, Intel®, AMD®, and NVIDIA®, we provide an accurate characterization of the real energy consumption of convolutional workloads. Our results highlight the limitations of MSR-based energy estimations, which systematically underestimate actual energy consumption by 10âĂŞ30% compared to socket-level measurements. This discrepancy demonstrates that socket-level instrumentation offers higher reliability and precision for energy benchmarking on modern CPUs. We also observe a consistent trade-off between inference latency and power consumption across all evaluated architectures. Overall, the ARM® Cortex-A78AE combined with an implicit GEMM convolution implementation achieves the best balance, reaching ResNet50v1.5 inference in 102 ms at 25.3 W on average, corresponding to 2.58 J. Future work will focus on defining new energy benchmarking standards that account for the growing limitations of MSRs, notably their reduced ability to expose DRAM energy consumption. We advocate for approaches that combine MSR data with socket-level measurements to enable more accurate and sustainable per-component energy estimation across evolving CPU architectures.

References

1. Abdelfattah, A., Haidar, A., Tomov, S., Dongarra, J.: Performance, design, and autotuning of batched GEMM for GPUs. In: Kunkel, J.M., Balaji, P., Dongarra, J. (eds.) High Performance Computing, pp. 21–38. Springer International Publishing, Cham (2016)
2. Carion, N., Massa, F., Synnaeve, G., Usunier, N., Kirillov, A., Zagoruyko, S.: End-to-end object detection with transformers. In: Computer Vision – ECCV 2020: 16th European Conference, Glasgow, UK, August 23–28, 2020, Proceedings, Part I, pp. 213–229. Springer-Verlag (2020)
3. Cassagne, A., Amiot, N., Bouyer, M.: Dalek: An unconventional and energy-aware heterogeneous cluster (2025). https://doi.org/10.48550/arXiv.2508.10481
4. Dolz, M.F., et al.: Performance–energy trade-offs of deep learning convolution algorithms on ARM processors. J. Supercomput. **79**(9), 9819–9836 (2023)
5. Dolz, M.F., Martinez, H., Alonso, P., Quintana-Orti, E.S.: Convolution operators for deep learning inference on the Fujitsu A64FX Processor. In: 2022 IEEE 34th International Symposium on Computer Architecture and High Performance Computing (SBAC-PAD) (2022)

6. Du, J.: Understanding of object detection based on CNN family and YOLO. J. Physics: Conf. Series **1004**(1) (2018)
7. Georganas, E., et al.: Anatomy of high-performance deep learning convolutions on SIMD architectures (2018). https://doi.org/10.48550/ARXIV.1808.05567
8. He, K., Zhang, X., Ren, S., Sun, J.: Deep residual learning for image recognition. In: 2016 IEEE Conference on Computer Vision and Pattern Recognition (CVPR), pp. 770–778 (2016)
9. Intel: OneAPI Deep Neural Network Library (OneDNN). https://oneapi-src.github.io/oneDNN/
10. Jay, M., Ostapenco, V., Lefevre, L., Trystram, D., Orgerie, A.C., Fichel, B.: An experimental comparison of software-based power meters: focus on CPU and GPU (2023). https://doi.org/10.1109/CCGrid57682.2023.00020
11. Khan, K., Hirki, M., Niemi, T., Nurminen, J., Ou, Z.: RAPL in action: Experiences in using RAPL for power measurements. ACM Trans. Modeling Performance Eval. Comput. Syst. (TOMPECS) **3** (2018). https://doi.org/10.1145/3177754
12. Khanam, R., Hussain, M.: YOLOv11: An overview of the key architectural enhancements (2024). https://doi.org/10.48550/arXiv.2410.17725
13. Kågström, B., Ling, P., van Loan, C.: GEMM-based level 3 BLAS: high-performance model implementations and performance evaluation benchmark. ACM Trans. Math. Softw. **24**(3), 268–302 (1998)
14. Krizhevsky, A., Sutskever, I., Hinton, G.E.: Imagenet classification with deep convolutional neural networks. Commun. ACM **60**(6), 84–90 (2017)
15. Lavin, A., Gray, S.: Fast algorithms for convolutional neural networks (2015). https://doi.org/10.48550/ARXIV.1509.09308
16. Lin, T.Y., et al.: Microsoft coco: common objects in context. In: Fleet, D., Pajdla, T., Schiele, B., Tuytelaars, T. (eds.) Computer Vision - ECCV 2014, pp. 740–755. Springer International Publishing, Cham (2014)
17. Lym, S., Lee, D., O'Connor, M., Chatterjee, N., Erez, M.: DeLTA: GPU Performance Model for Deep Learning Applications with In-depth Memory System Traffic Analysis. In: 2019 IEEE International Symposium on Performance Analysis of Systems and Software (ISPASS), pp. 293–303 (2019)
18. Ma, Y., et al.: Revisiting DETR pre-training for object detection (2023). https://arxiv.org/abs/2308.01300
19. Meng, L., Brothers, J.: Efficient winograd convolution via integer arithmetic (2019). https://doi.org/10.48550/ARXIV.1901.01965
20. Park, H., Kim, D., Ahn, J., Yoo, S.: Zero and data reuse-aware fast convolution for deep neural networks on GPU. In: Proceedings of the Eleventh IEEE/ACM/IFIP International Conference on Hardware/Software Codesign and System Synthesis. CODES '16, Association for Computing Machinery, New York, NY, USA (2016). https://doi.org/10.1145/2968456.2968476
21. Santana, A.d.L., Armejach, A., Casas, M.: Efficient direct convolution using long SIMD instructions. In: Proceedings of the 28th ACM SIGPLAN Annual Symposium on Principles and Practice of Parallel Programming. PPoPP '23, ACM (2023)
22. Sze, V., Chen, Y.H., Yang, T.J., Emer, J.: Efficient Processing of Deep Neural Networks: A Tutorial and Survey. Tech. rep. (2017). https://doi.org/10.48550/arXiv.1703.09039
23. Wei, H., Liu, E., Zhao, Y., Yu, H.: Efficient non-fused winograd on GPUs. In: Advances in Computer Graphics. pp. 411–418. Springer International Publishing, Cham (2020)
24. Winograd, S.: Arithmetic complexity of computations. Soc. Industrial Appl. Math. (1980). https://doi.org/10.1137/1.9781611970364

25. Yan, D., Wang, W., Chu, X.: Optimizing batched winograd convolution on GPUs. In: Proceedings of the 25th ACM SIGPLAN Symposium on Principles and Practice of Parallel Programming, pp. 32–44. New York, NY, USA (2020)
26. Zhang, J., Franchetti, F., Low, T.M.: High performance zero-memory overhead direct convolutions (2018). https://doi.org/10.48550/ARXIV.1809.10170
27. Zhou, Y., et al.: Characterizing and demystifying the implicit convolution algorithm on commercial matrix-multiplication accelerators (2021). https://doi.org/10.48550/ARXIV.2110.03901
28. Zhuo, S., et al.: SCL-YOLOv11: a lightweight object detection network for low-illumination environments. IEEE Access **13**, 47653–47662 (2025)

C3M-ViT: A Configurable Matrix Multiplication Module for Energy-Efficient Vision Transformer Accelerator

Thomas Jacquemin, Hana Krichene(✉), and Benoit Tain

Université Paris-Saclay, CEA, List, F-91120 Palaiseau, France
{thomas.jacquemin,hana.krichene,benoit.tain}@cea.fr

Abstract. The rapid progression of Artificial Intelligence has made Vision Transformers (ViTs) crucial for computer vision, offering accuracy comparable to or exceeding that of CNNs. However, ViTs present implementation challenges on hardware platforms, requiring the processing of larger matrices, leading to substantial parameter counts and demanding memory accesses. In response to these challenges, this paper introduces a Configurable Matrix Multiplication Module C3M-ViT. It employs a highly parallel temporal structure based on clusters of Processing Elements (PEs) organized in a two-dimensional matrix. By leveraging a tiling methodology and an optimized dataflow, the C3M-ViT achieves efficient storage without the need for data rearrangement enabling uniform handling of both input and output layers, while maintaining a high PE utilization rate. Furthermore, the C3M-ViT's weight loading strategy significantly reduces memory footprint by eliminating the necessity to store entire weight matrices. This paper present a 5×4 cluster matrix configuration with 1,446 parallel PEs, delivering 77.6 $GOPs/W$ energy efficiency, which is approximately three times higher than comparable state-of-the-art accelerators under similar conditions.

1 Introduction

Artificial Intelligence (AI) has progressed rapidly in recent years, with deep neural networks driving major advances in computer vision, language processing and autonomous systems. A major technological shift occurred with the introduction of the Transformer network architecture [9]. Specifically, the ViT model [3], proposed in 2021, adapted the Transformer's encoder for image classification tasks. By processing embedded patches through the self-attention mechanism, ViTs offer accuracy comparable to or exceeding that of Convolutional Neural Networks (CNNs) on standard vision benchmarks. ViTs overcome the limitations of traditional CNNs in maintaining feature dependencies over large networks and offer inherently better possibilities for parallelizing calculations.

Despite these advances, implementing ViT models efficiently on hardware platforms presents unique challenges. ViTs require the processing of significantly larger matrices than CNNs, leading to a substantial parameter count. These

M. Kowalczyk and C. Monière (Eds.): DASIP 2026, LNCS 16512, pp. 95–106, 2026.
https://doi.org/10.1007/978-3-032-23871-9_8

larger matrices require intense and regular calculations and demand numerous memory accesses. Critically, non-computational processes, such as matrix handling and rearrangement, also consume time during large matrix multiplications.

FPGAs have become one of the favored platform for ViT accelerators due to their reconfigurability and balance between computational performance and energy efficiency. However, limited on-chip memory in FPGA makes efficient dataflow and precise memory access management essential to extract maximum platform performance.

This paper presents the design and development of C3M-ViT a hardware acceleration module dedicated to matrix multiplication. This operation accounts for over 95% of the total computation performed during ViT inference. C3M-ViT aims to optimize energy efficiency by accelerating calculations using a highly parallel architecture based on Processing Elements (PEs) and reducing the volume of accesses to internal module memory and off-chip memory. Furthermore, C3M-ViT proposes to reduce on-chip memory. And ensures the module remains parametric and adaptable to various ViT architectures and diverse applications. The rest of this paper is organized as follows : Sect. 2 reviews related works and background. Section 3 describes C3M-ViT and overall ViT accelerator architecture. Section 4 presents the experimental results, providing insights into its performances. Finally, Sect. 5 concludes the paper by summarizing the key findings.

2 Related Work and Background

2.1 ViT Model

The ViT architecture is based on the Transformer's encoder, consisting primarily of Multi Self-Attention (MSA) and Multi-Layer Perceptron (MLP) groups. Linear functions, which are all derived from matrix multiplications, are ubiquitous throughout the network and account for nearly 95% of the total calculations. These operations include the generation of Query (Q), Key (K), and Value (V) matrices, the core self-attention calculations such as $Q \times K^T$ and $Softmax(Q \times K^T) \times V$, and the Fully connected (FC) operations in the MLP group. For instance, FC operations involve multiplying the input neuron layer by a weight set, effectively performing matrix computation. The ViT-tiny model introduce in [8] was selected for implementation because, despite being one of the smallest available with only 5.8M parameters, its substantial matrix sizes are still quite demanding in terms of memory for embedded systems.

2.2 Matrix Multiplication Accelerator Paradigms

Hardware acceleration generally employs either pipeline architectures [4,10] or temporal architectures [5,6]. Pipeline architectures offer high bandwidth but demand substantial hardware resources. Conversely, temporal architectures, which utilize a central, reusable computing unit, most often a systolic array, are more resource-efficient and better suited for ViTs, especially in embedded systems, due to the similarity of their linear computations. Various methods for implementing these temporal accelerators exist, often revolving around a 2D matrix of PEs, called clusters.

A large proportion of ViT accelerators employ clusters with Weight Stationary (WS) dataflows [5]. This type of dataflow helps minimize energy consumption associated with reading weights. This dataflow is widely adopted in CNN accelerators, significantly reducing memory access requirements through the high weight-sharing characteristic of CNN kernels.

Despite the benefits of these dataflow optimizations, current ViT accelerators often parallelize calculations and data loading but still face the challenge of storing the entire weight matrix in the accelerator's memory. This results in high memory requirements and limits scalability, especially for large-scale models.

2.3 Positioning

In this work, we introduce a temporal configurable Matrix Multiplication Module implemented on FPGA, specifically adapted for Vision Transformer (ViT) networks. The C3M-ViT architecture is designed to achieve low operational energy costs and minimal resource requirements, particularly in terms of memory footprint, while providing both flexibility and scalability. Our contributions include:

- An optimized Output Stationary dataflow architecture with 8-bit quantization enabling the execution of all linear functions of ViT.
- Configurable computation units and memory size, employing a generic calculation decomposition method that enables hardware adaptation to various applications while providing flexibility and scalability.
- A memory management strategy enabling uniform handling of both input and output layers, facilitating seamless end-to-end processing without temporal overhead.

3 Configurable Matrix Multiplication Accelerator

C3M-ViT is designed to maximize flexibility, scalability, and reduce PE inactivity, for ViT model inference. The C3M-ViT architecture focuses on optimizing the linear functions derived from massive matrix multiplications. The overall ViT architecture is illustrated in Fig. 1. Data are stored in two types of buffers: I/O and In, and data management is orchestred by a Global Controller to move data from buffer to 2D cluster matrix. The C3M-ViT is interfaced with a Non-linear Unit (NLU) [7] that performs LayerNorm, Softmax, and GeLU operations. This interconnection enables the implementation of a complete ViT accelerator.

The architecture adopts a generic calculation decomposition method that provides flexibility and scalability. This approach ensures that the accelerator can efficiently and easly adapted to a wide and varied range of matrix sizes. This adaptability allows the C3M-ViT to scale effectively, making it suitable for a variety of ViT models. A key objective of this design is to store data without requiring rearrangement, simplifying memory management while enabling

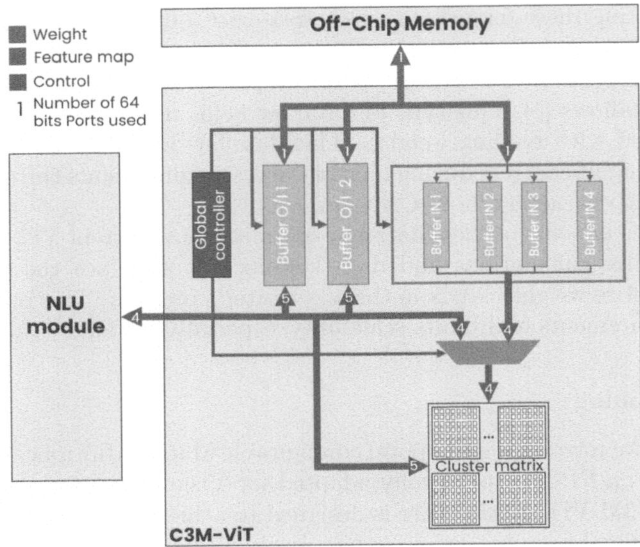

Fig. 1. ViT Accelerator overview.

uniform handling of both input and output layers. This standardization of data layer formats enhances the end-to-end processing workflow without temporal overhead.

3.1 A Modular and Configurable Architecture

The C3M-ViT architecture features a modular and parametric design, allowing for optimized configurations tailored to specific computational requirements. The core computation is handled by a two-dimensional matrix of PE clusters. This modularity ensures that the hardware can be easily adapted to different ViT configurations by modifying number of cluster. For the ViT model, an intensive comparative study concluded that the most optimal configurations are in the shape of a $5 \times x$ cluster matrix. For the ViT-tiny model, a configuration of a 5×4 cluster matrix was chosen, this configuration minimize PE inactivity for a ViT-tiny application. This specific configuration, visualized in Fig. 2, contains 20 clusters of 64 PEs, totaling 1280 parallel computational units.

Data storage and flow are managed through specialized buffers. Buffers I/O (Input/Output) are the main storage for Feature Maps (FM) and results. These buffers are substantially sized (1.632 Mbit capacity, implemented using 51 Block RAMs (BRAMs) used in Simple Dual-Port (SDP) mode) and include multiple read/write ports equal to the maximum number of rows or columns in the cluster matrix. Each port can transfer 128 bits, 64 bits dedicated to writing and 64 bits to reading. The addressing scheme for these buffers ensures parallel access by dividing the data address into a 6-bit BRAM index and a 9-bit posi-

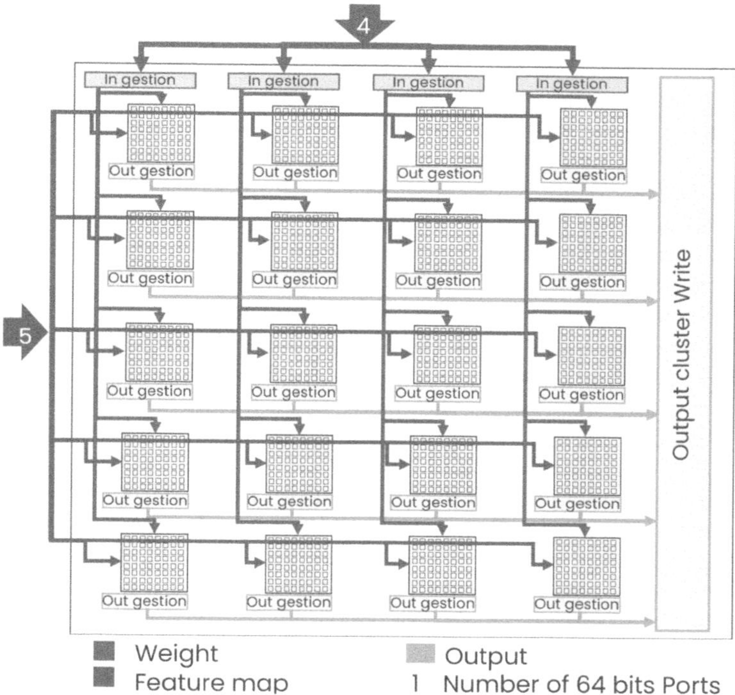

Fig. 2. Architecture of cluster matrix (5 × 4).

tion index within the BRAM, preventing conflicts when multiple ports access data simultaneously. In 5 × 4 configuration, by sending addresses with different BRAM indices, it ensures that each port accesses a different BRAM, enabling a maximum 640-bit read/write simultaneously.

In contrast, the Buffers In store the weights (W). These are smaller (128 Kbit, using 4 BRAMs in SDP mode) and are designed to hold only two vertical sub-matrices of weights. This approach allows C3M-ViT to avoid storing entire weight matrices within the accelerator, thereby minimizing memory footprint. The single read port of Buffers In connects to an entire column of clusters, enabling simultaneous diffusion of the weight values.

The entire module is coordinated by a Global Controller, which supervises and manages data flow by generating necessary Buffer I/O and In addresses. The controller's behavior is configured by a 42-bit configuration word, which defines matrix dimensions (M×D and D×N), data buffers source and destination, layer type and the number of heads. Since the data are stored without rearrangement, it is possible to construct the 15-bit address of all the data from their position in the matrix. Thus, the data addresses are constructed using a 7-bit column index positioned on the least significant bits (LSB) and an 8-bit row index placed on the most significant bits (MSB) of the address.

3.2 Dataflow and Control

The C3M-ViT employs adaptive control and dataflow techniques, utilizing methods such as matrix tiling, and multi-broadcasting in order to reuse data and reduce memory usage and accesses, thereby improving energy efficiency. Each PE cluster (an 8×8 PE matrix) implements an Output Stationary dataflow. This type of dataflow minimizes energies consumption of the writing and reading of partial products and allow a data process without rearrangement. Partial products remain in the 16-bit register bank in the PE until the final result is accumulated. This allows us to avoid having dedicated memory for storing partial products. This memory is often composed of registers storing data with a higher quantization than the FM and W, and it size is highly dependent on the size of the application's matrix. Thus, this dataflow allows us to maintain the same number of registers within the cluster matrix, regardless of the sizes of the FM and W matrix. The PE, the core unit, performs a MAC operation in a single clock cycle by leveraging dedicated DSP Slices.

The massive matrix multiplication (M×D by D×N) is decomposed by tiling. Each cluster receives an $8 \times D$ horizontal FM sub-matrix and a $D \times 8$ vertical W sub-matrix, producing an 8×8 block of the output matrix. The dataflow illustrating this decomposition is shown in Fig. 3. This tiling is adaptive to resources instanced by the module.

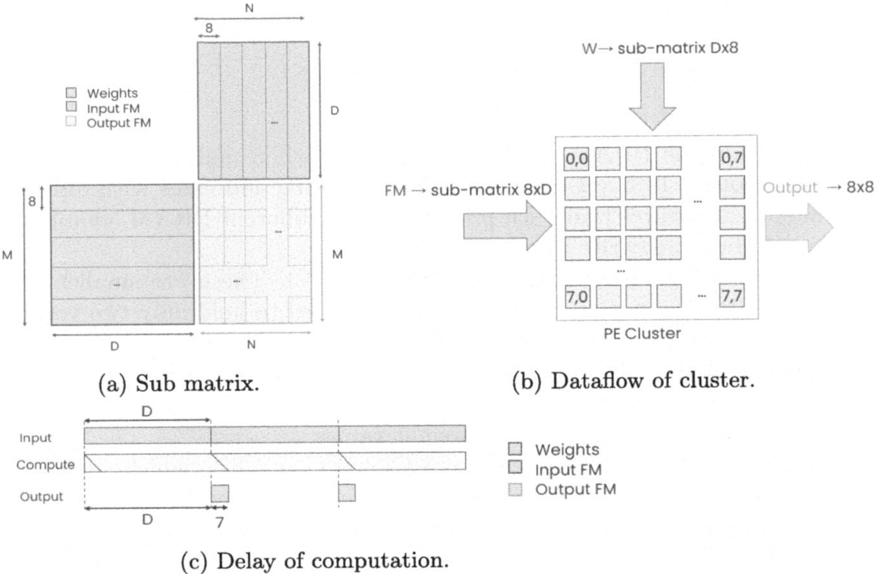

(a) Sub matrix.

(b) Dataflow of cluster.

(c) Delay of computation.

Fig. 3. Dataflow and sub matrix.

Within the cluster, there is a propagation of weight from one PE to the one directly below it, and a diffusion of feature maps (FM) across an entire row of the cluster. At each cycle, the cluster receives a horizontal vector of 8 weight values and a vertical vector of 8 FM values, totalizing 64 bits per vector.

Since FM data is stored in the buffers by rows of 8 values but required by the clusters in columns of 8 values, the Input Gestion sub-module transforms row-based memory accesses into the required column accesses. It utilizes an 8×8 register matrix to receive a line of 8 values per clock cycle and transmits a column of 8 values simultaneously to the clusters. This data transformation process, which involves shifting the data matrix to consume a new value every clock cycle, is shown in Fig. 5. This mechanism ensures the necessary data bandwidth (8 FM values per cycle) is met while maintaining alignment in the memory accesses. This sub-module ensures a readability of data stored without rearrangement.

Accessing the data also follows specific strategy depending on the matrix orientation. Reading a vertical sub-matrix $D \times 8$ is straightforward, achieved by incrementing the line index from 0 to $D - 1$. Reading a horizontal sub-matrix $8 \times D$ is more complex and involves 8 accesses while incrementing the line index before incrementing the column index, leveraging the Input Gestion sub-module to handle the rearrangement.

The architecture supports two dataflow scenarios: calculations with weight matrices $FM \times W$ (mainly in MLP) and calculations without weight matrices $A \times B$ (mainly in MSA). When the calculation involves weights, they are loaded into Buffers In and then provided to the vertical input of the multiplication module, with FM being provided on the horizontal input. In the case of a calculation without weights, the two matrices are present in the two I/O buffers. Thus, the buffer I/O containing matrix B will be connected to the vertical input of the module, and the buffer I/O containing matrix A will be connected to the horizontal input.

3.3 Memory and Data Management

All input and output data are stored in Buffers I/O without rearrangement, enabling uniform and consistent data organization across all layers. This is achieved through pre-writing rearrangement techniques that overlap communication and computation operations. By enabling simultaneous data loading from off-chip memory and parallel processing, total execution time is reduced, maximizing throughput and minimizing idle periods. This approach also masks non-computational functions, such as transposition during transfers.

A crucial component in achieving this overlap is the utilization of Buffers In, which are sized to hold two consecutive $D \times 8$ vertical sub-matrices of weights. While the current layer is being processed using weights from one part of the buffer In, the subsequent set of weights can be loaded from external memory into the other part of the buffer In (Fig. 4). This parallelism effectively masks the latency associated with external memory access and ensures that calculation units remain continuously utilized. For the 5×4 configuration of cluster matrix, the loading time (4 * D clock cycles) is less than the calculation time (5 * D clock cycles), which means that the computation limits the overall speed.

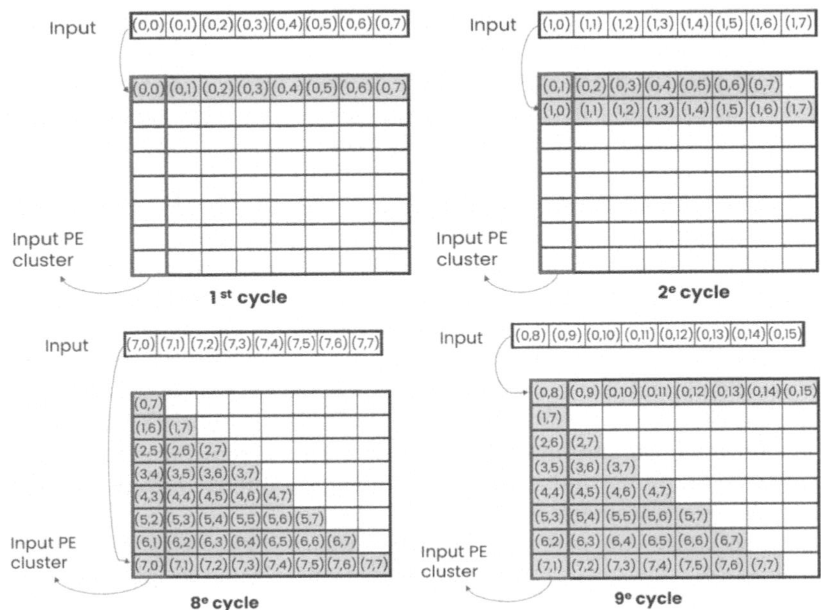

Fig. 4. Buffer In dataflow for a C3M-ViT in configuration 5×4.

The architecture also manages the bottleneck created when all 20 clusters finish calculating simultaneously after D clock cycles. The Output Cluster Write sub-module is responsible for collecting these 20 product lines over 8 clock cycles and perform on-the-fly matrix transposition if needed. By transmitting the results as columns instead of rows, the sub-module can store the transposed matrix in the I/O buffers with no added latency. The global controller constructs transpose addresses by inverting row and column indices. Additionally, thanks to the data storage policy without rearrangement, matrices are stored concatenated during writes.

4 Experimental Results

The overall ViT accelerator was implemented on the ZCU102 FPGA board. This platform was chosen because of its frequent use in the literature and its substantial integrated resources, notably 912 BRAMs and 2520 DSP Slices. The model selected for testing is the ViT-tiny as presented in [8]. Chosen for its compact nature with only 5.8 million parameters, it minimizes hardware and energy constraints, although its largest layers define the critical dimensions of the accelerator. The development was carried out in VHDL, and validation tests were performed in the Vivado environment.

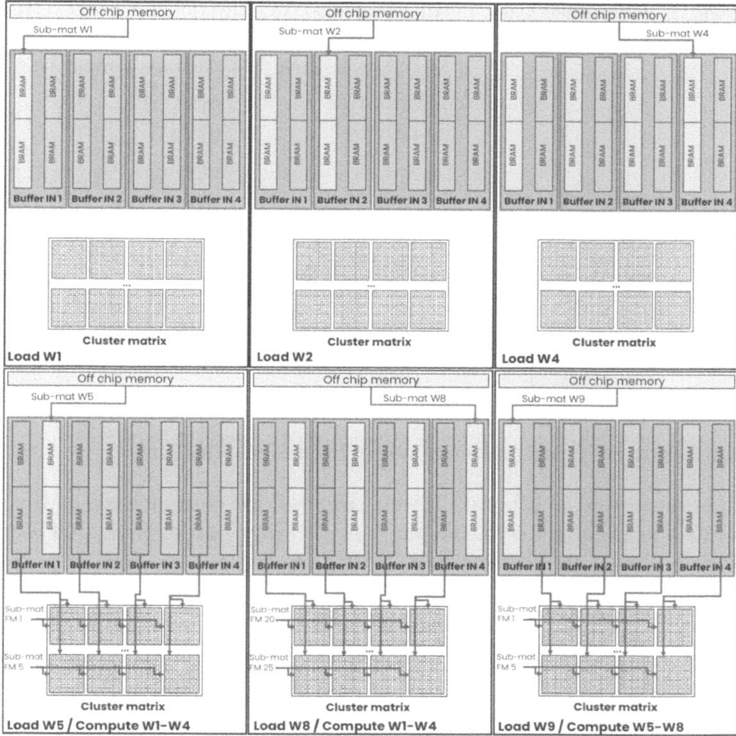

Fig. 5. Transformation of row to column access done in sub-module In Gestion.

4.1 Implementation and Resource Utilization

The design phase resulted in the implementation and analysis of two primary configurations of the C3M-ViT cluster matrix architecture: a 5 × 4 cluster matrix configuration and a 5 × 6 cluster matrix configuration. Resource utilization estimates (Table 1) obtained after placement and routing in Vivado confirmed that both implemented configuration efficiently utilize the available resources while maintaining a relatively low occupancy. For the C3M-ViT 5 × 4 configuration, 1,280 of the available 2,520 DSPs were used. The C3M-ViT 5 × 6 configuration, designed for higher parallelism, required significantly more, using 1,920 DSP slices.

Crucially, BRAM usage is extremely low for both designs, reflecting the core architectural strategy of overlapping computation and weight loading to reduce storage requirements. The 5 × 4 and 5 × 6 configurations use 118 BRAMs and 126 BRAMs, respectively, with the 5 × 6 configuration having more Buffer In. This ensures a low memory footprint of 3.7 Mbits and 4 Mbits for a ViT-tiny application.

Table 1. Resources Used by Implemented ViT accelerator on ZCU102

Type	Available	ViT Accelerator configuration 5 × 4		ViT Accelerator configuration 5 × 6	
		C3M-ViT 5 × 4	Total	C3M-ViT 5 × 6	Total
kLUT	274	75	83	103	111
kLUTRAM	144	0	1.2	0	1.2
kFF	548	44	49	65	72
BRAM	912	118	118	126	126
DSP	2,520	1,280	1 446	1,920	2 166

4.2 Performance Comparison with State-of-the-Art

Accelerators presented in [2] and [5] were selected for comparison, as both implement designs specifically for executing the ViT-tiny model with INT8 quantization. Additionally, both utilize the ZCU102 board as their hardware platform. These references provide a relevant benchmark for evaluating the performance of our proposed C3M-ViT configuration in the context of ViT-tiny inference. In terms of overall execution time, our proposed 5 × 4 configuration, operating at 100 MHz, exhibits a higher latency of 10.050 ms compared to the architectures presented in [5] (4.077 ms) and [2] (3.690 ms). However, this difference is mainly attributed to the lower operating frequency compared to [5] (300 MHz) and [2] (150 MHz). When analyzing the computational efficiency based on the number of clock cycles required for complete ViT inference, the proposed 5 × 4 configuration is superior. It achieves 2,392 *Ops/cycle* and realizes inference in 1,050k *cycles*, while the method in [5] requires 1,200k *cycles* for an inference while performing 2,093 *Ops/cycle*. This efficiency gain stems from the cluster design, thus reducing layer latency. This superior cycle count is achieved even though [5] utilizes the DI8MAC [1] method, which performs two MAC operations per DSP slice and theoretically 4 Operations per cycle per DSP. Similarly, [2] achieves inference in 553k *cycles*, but [2] use a pruned model that greatly reduces the number of operations needed, reaching only 1,356 *ops/cycle* (Table 2).

Table 2. Comparison of Our ViT Accelerator Performances with Related Works.

Architecture	kLUT	kFF	BRAM	DSP	Latency (*ms*)	Throughput (*GOPs*)	Frequency (*MHz*)
ViT accelerator configuration 5 × 4 [ours]	83	49	118	1,446	10.050	249.9	100
ViT accelerator configuration 5 × 6 [ours]	158	72	126	2,166	8.345	301.0	100
Group vector SA [5]	114	168	648	1,268	4.077	616.1	300
Heat ViT [2]	137	126	355	1,968	3.690	203.0	150

A major architectural advantage is evident in the resource footprint, particularly memory. The 5×4 configuration requires only 3.7 Mbits of memory, which is drastically lower than the 20.7 Mbits used by [5] and the 11.3 Mbits used by [2] (Table 3).

Table 3. Comparison of Our ViT Accelerator Energy Efficiency with Related Works

Architecture	Power(W)*	Energy Efficiency ($GOPs/W$)
ViT accelerator configuration 5×4[ours]	3.08	77.6
ViT accelerator configuration 5×6[ours]	5.34	56.3
Group vector SA [5]	29.60	20.8
HeatViT [2]	9.45	21.5

*Dynamic Power consumption of FPGA

Finally, the overall architecture demonstrates superior energy efficiency (Table 3). The 5×4 configuration achieves 77.6 $GOPs/W$, which is approximately 3 times superior to the efficiency of [5] (47.56 $GOPs/W$) and [2] (21.48 $GOPs/W$). This significant efficiency gain is explained by the very low dynamic power consumption of 3.081 W for the 5×4 configuration, resulting from the optimization of logic and memory resource utilization. The 5×6 configuration, while more powerful, achieved a lower efficiency of 56.33 $GOPs/W$.

5 Conclusion

The developed C3M-ViT architecture provides a compelling solution for low-energy, memory-efficient matrix multiplication acceleration for ViTs application which prioritize a low memory footprint and high energy efficiency. The architectural strengths include parallelization of computation and weight loading, as well as the tiling method, which reduces the memory footprint and mitigates latency by reducing buffer sizes. The design implements a dataflow that reduces the energy cost associated with reading and writing partial products and limits the energy cost of reading inputs and weights through a broadcasting strategy. Furthermore, the module minimizes non-calculatory matrix operations, such as transposition and concatenation, with a no rearrangement strategy, which save processing time.

Experimental results confirm the efficiency of the approach when implemented on the ZCU102 FPGA using the ViT-tiny model. The ViT accelerator in 5×4 configuration achieved an energy efficiency of 77.6 $GOPs/W$, which is approximately three times superior to comparable state-of-the-art accelerators. Crucially, the architectural focus on reducing storage requirements resulted in a dramatically small memory footprint of only 3.7 Mbits for the ViT-tiny application. This combination of effective parallelization, minimized data rearrangement, and configurable design ensures flexibility and scalability for various matrix sizes and computational requirements, making it an efficient solution for modern AI workloads.

References

1. 8-bit dot-product acceleration. Xilinx Inc., San Jose, CA, USA, Xilinx White Paper WP487 v.1.0 (2017)
2. Dong, P., et al.: Heatvit: hardware-efficient adaptive token pruning for vision transformers (2023)
3. Dosovitskiy, A., et al.: An image is worth 16x16 words: Transformers for image recognition at scale (2021)
4. Guo, Q., Wan, J., Xu, S., Li, M., Wang, Y.: Hg-pipe: Vision transformer acceleration with hybrid-grained pipeline (2024)
5. Huang, M., Luo, J., Ding, C., Wei, Z., Huang, S., Yu, H.: An integer-only and group-vector systolic accelerator for efficiently mapping vision transformer on edge. IEEE Trans. Circuits Syst. I Regul. Pap. **70**(12), 5289–5301 (2023)
6. Liu, Z., Li, G., Cheng, J.: Hardware acceleration of fully quantized BERT for efficient natural language processing (2021)
7. Nguyen, L.N.H., Krichene, H.: Efficient FPGA implementation of vit non-linear functions. In: Lorandel, J., Kamaleldin, A. (eds.) Design and Architecture for Signal and Image Processing, pp. 16–27. Springer Nature Switzerland, Cham (2025)
8. Touvron, H., Cord, M., Douze, M., Massa, F., Sablayrolles, A., Jégou, H.: Training data-efficient image transformers & distillation through attention (2021). https://arxiv.org/abs/2012.12877
9. Vaswani, A., et al.: Attention is all you need (2023)
10. Zhuang, J., et al.: Ssr: spatial sequential hybrid architecture for latency throughput tradeoff in transformer acceleration. In: Proceedings of the 2024 ACM/SIGDA International Symposium on Field Programmable Gate Arrays, pp. 55–66. FPGA '24, ACM (2024)

Comparative Benchmark of Scaled-YOLOv4 on Graphcore IPU and NVIDIA GPU

Arjan Siddhpura, S.-Kazem Shekofteh(✉), and Holger Fröning

Hardware and Artificial Intelligence (HAWAII) Lab, Heidelberg University, Heidelberg, Germany
arjan.siddhpura@stud.uni-heidelberg.de,
{kazem.shekofteh,holger.froening}@ziti.uni-heidelberg.de

Abstract. The performance of deep learning models is heavily coupled to hardware, with SIMT-based GPUs being the de facto standard. Novel architectures like the Graphcore Intelligent Processing Unit (IPU), with its Multiple Instruction, Multiple Data (MIMD) design and distributed on-chip SRAM, offer a different paradigm. However, direct performance comparisons for complex, state-of-the-art computer vision models are sparse. This paper presents a comprehensive performance benchmark of the Scaled-YOLOv4-P5 object detection model on a Graphcore GC200 IPU against a comparable NVIDIA A30 GPU. We investigate the performance trade-offs by analyzing inference latency and throughput while varying image size, batch size, and floating-point precision. Our findings reveal a stark performance trade-off. The IPU excels in low-latency scenarios, delivering a 6.56 ms inference time at batch=1 (896 px), nearly 4x faster than the GPU's 26.17 ms. Conversely, the GPU's SIMT architecture scales near-linearly for high-throughput, while the IPU is severely memory-constrained. The IPU failed to compile at batch=2 for the native 896px resolution, limited by its ~900 MB on-chip SRAM. In contrast, the GPU's 24 GiB HBM2 memory handled batches of ≥64 at the same resolution. Furthermore, the IPU's Ahead-of-Time compilation incurs a major overhead: a full benchmark run at 896 px took 382.79 s on the IPU versus just 15.56 s on the GPU, with 75–88% of the IPU's time spent on compilation alone.

Keywords: Object Detection · Graphcore IPU · YOLOv4 · Performance Benchmarking · Hardware Accelerators

1 Introduction

The performance of deep learning is inextricably linked to the evolution of specialized hardware, a dependence famously described as the "Hardware Lottery" [9]. For years, this field has been dominated by Graphics Processing Units (GPUs), whose Single Instruction, Multiple Threads (SIMT) architectures are

M. Kowalczyk and C. Monière (Eds.): DASIP 2026, LNCS 16512, pp. 107–118, 2026.
https://doi.org/10.1007/978-3-032-23871-9_9

serendipitously efficient at processing the dense matrix multiplications that form the backbone of modern computer vision models. This SIMT design thrives on massive data parallelism, achieved by processing large batches. However, this is a disadvantage for applications like real-time object detection, which demand the lowest possible latency on a single input (batch size 1). This mismatch has spurred the development of novel AI accelerators built on different paradigms like the Graphcore Intelligent Processing Unit (IPU), a massively parallel processor built on a Multiple Instruction, Multiple Data (MIMD) design with ~900 MB of distributed on-chip SRAM. Such an architecture is ideal for workloads with fine-grained, irregular parallelism and is designed to deliver low latency by co-locating memory and compute.

This architectural dichotomy however lacks a direct comparison for state-of-the-art (SOTA) vision models. While vendor-published benchmarks for the IPU claim significant inference speedups—including a 2.2x latency and 4.7x throughput improvement on Resnet-50 [8], a 15x latency and 14x throughput improvement on EfficientNet-B0 [18] and a 24x latency and 7x throughput on ResNeXt-101 [22], these claims [7] require independent validation. Other studies often compare hardware from different generations (e.g., first-generation IPU MK1 (GC2) vs. NVIDIA V100) [1], different power envelopes, or different target workloads. These studies often focus on tasks such as matrix multiplication [10, 15] or vision tasks like text detection [17] and cosmology image generation [1].

What is missing is an empirical benchmark that compares these distinct architectural paradigms—the SRAM-centric MIMD approach versus the HBM-centric SIMT approach—on a complex, irregular workload. While newer hardware generations have emerged (e.g., NVIDIA H100, Graphcore Bow), the A30 and GC200 represent a stable, accessible, and power-comparable (150W–165W) mid-range class widely used in inference. More importantly, the fundamental trade-offs they exhibit—latency minimization via SRAM versus throughput maximization via HBM—remain the primary bottleneck determinants in modern accelerator design. To this end, we benchmark the 70.8-million-parameter Scaled-YOLOv4-P5 [20].

The objective of this work is to fill this specific gap. We present a rigorous, quantitative performance benchmark of the Scaled-YOLOv4-P5 model on a Graphcore GC200 IPU versus a power-comparable NVIDIA A30 GPU. We analyze the architectural trade-offs by measuring end-to-end inference latency and throughput while varying input image size, batch size, and floating-point precision.

Our primary contributions are:

1. Confirming the feasibility along with a benchmark of the ~70.8M-parameter Scaled-YOLOv4-P5 model on a GC200 IPU and A30 GPU, quantifying the IPU's significant low-latency advantage (approx. 4x speedup) in real-time scenarios (batch size 1).
2. Identification of the IPU's critical performance bottlenecks: the ~900 MB on-chip SRAM memory wall which prevents compilation of larger batch sizes, and

the significant overhead incurred by its Ahead-of-Time (AOT) compilation model.

This work provides empirical data for system designers, demonstrating the fundamental performance trade-off between the MIMD/SRAM (specialization) and SIMT/HBM (generalization) architectural designs.

The code for this benchmark is available online.

2 Background and Related Work

2.1 Object Detection Model

The YOLO family transformed object detection by treating the problem as a single regression task, enabling real-time performance. From its multiple iterations, we selected the Scaled-YOLOv4-P5 for our benchmark. While newer models exist, this model provides a representative benchmark for hardware. Its architecture is not just a simple, sequential CNN; it features a complex CSP-Darknet53 [21] backbone and a Path Aggregation Network (PANet) [13] neck. The backbone consists of dense convolutional layers, which align well with the high data parallelism of the GPU's SIMT architecture. Conversely, the PANet, with its top-down and bottom-up feature-fusion pathways, creates an irregular, graph-like computation flow favoring the IPU's MIMD flexibility. This model thus serves as a stress test for the competing SIMT and MIMD execution models.

Previous research also provides a reference for the performance of YOLOv4 variants on GPU hardware. For instance, the YOLOv4-tiny model achieved 371 FPS at 21.7% AP on an NVIDIA GTX 1080 Ti and 443 FPS on an NVIDIA RTX 2080 Ti [20]. The larger YOLOv4-P5 model, in turn, achieved 41–43 FPS on an NVIDIA Tesla V100 GPU [20], matching our results.

2.2 Hardware Accelerators

We chose two accelerators based on similar power consumption and pricing. Table 1 provides an overview of both the hardware platforms used.

The NVIDIA A30 GPU is a data-center accelerator based on the Single Instruction, Multiple Threads (SIMT) execution model. It comprises 56 Streaming Multiprocessors (SMs), each executing the same instruction on many data points in parallel. This throughput-oriented design's performance relies on latency hiding via a 24 GiB pool of off-chip HBM2 memory. Given the SIMT design, workloads characterized by irregular data access, control flow divergence or sparsity result in performance penalties on the GPU. Despite having a large memory capacity, the GPU is also limited by its memory bandwidth when the time spent transferring data outweighs the computational speedup.

The Graphcore GC200 IPU is a many-core MIMD processor which addresses such limitations. It is a latency-oriented design, which contains 1,472 independent "tiles", each a full processor core with 624 KiB of dedicated SRAM. This totals ~900 MB of on-chip memory, accessible at an aggregated 47.5 TB/s—more than 50x that of the A30's HBM2.

The Poplar SDK acts as an Ahead-of-Time (AOT) compiler. It analyzes the entire YOLOv4 compute graph, partitions it, and maps each operation (e.g. a specific convolution) and its required data (weights, activations) to a specific tile before execution, including a static, deterministic allocation for the distributed SRAM to minimize latency from caching, swapping, or prefetching. However, the constrained 624 KiB per tile leads to resource contention among instruction code, model weights and input data; for instance, the largest square matrix multiplication feasible on the GC200 IPU was limited to a side length of 3,584 elements (single-precision), utilizing only 17% of available ~900 MB of In-Processor Memory [15]. This memory bottleneck extends beyond matrix operations, though subsequent research on butterfly factorization [14,16] has shown that structured sparsity can yield up to 98.5% compression, offering potential strategies to alleviate constraints for large-scale models like YOLOv4.

At runtime, the entire model is resident on-chip when using a single IPU and no streaming memory. As input images arrive, the chip executes in a Bulk Synchronous Parallel (BSP)-style [19]: (1) all 1,472 tiles compute in parallel, (2) they all synchronize, (3) they exchange necessary data with other tiles and then repeat. This architecture enables significant performance advantages over GPUs in certain tasks; for example, the IPU achieved up to 100x speedup in Breadth-First-Search (BFS) execution time over an NVIDIA A100 GPU for large graphs with up to 1 million edges [3].

Benchmarks specific to YOLOv4 on IPU systems further underscore these benefits: the model attained 924 FPS with a latency of 6.49 ms for 896 px images at a batch size of 4 (half-precision) on the Open Graph Benchmark (OGB) [4], while the Graphcore Bow-2000 system delivered 903 FPS compared to 122 FPS on an NVIDIA A100 under similar conditions, representing a throughput increase of more than 7 times [2]. Recent benchmarking suites like CARAML [11] also provide a broad, automated assessment of energy and performance for training workloads (e.g., ResNet50, LLMs) across NVIDIA, AMD, and Graphcore hardware. Their findings on the GC200 corroborate the "SRAM wall" limitation, noting that performance peaks within SRAM limits but struggles to scale with global batch size for training [11]. Our work complements such broad suites by conducting a deep-dive study of inference specifically for Scaled-YOLOv4's irregular graph structure.

3 Methodology

We designed an experiment to measure the inference performance of the identical Scaled-YOLOv4-P5 model on both hardware platforms while varying key parameters. For our experiments, we utilized a Graphcore IPU-M2000 blade (containing 4x GC200 processors), isolating a single GC200 chip for the benchmark to ensure a direct 1-to-1 comparison with the single NVIDIA A30 PCIe card. While both architectures support multi-chip scaling (via IPU-Link and

NVLink, respectively), restricting the benchmark to a single chip allows us to isolate the performance characteristics of the compute-memory subsystem (SRAM vs. HBM) without the confounding factors of interconnect bandwidth and the respective synchronization overheads.

Table 1. Specification comparison of the environment setup used for the benchmarking

Chip	Graphcore GC200 IPU	NVIDIA A30 GPU
Number of Parallel Cores	1472	3584
Memory per Parallel Unit [KiB]	624	192
Number of Threads	8832	229376
Total SRAM [MiB]	918	34.75 (L1 + L2)
Total DRAM [GiB]	256	24
DRAM Bandwidth [GB/s]	20	933
Clock Frequency [GHz]	1.33	1.44
FP32 Peak Compute [TFLOP/s]	62.5	10.3
Power Consumption [W]	150	165
Inter-Chip-Bandwidth [GB/s]	350	200
Host-Device Memory Bandwidth[a] [GB/s]	14.55	31.5
Device-Host Memory Bandwidth[a] [GB/s]	6.33	31.5
Software Stack	PopTorch 3.40.0	PyTorch 2.4.1
CPU Host Processor	2x AMD EPYC 7443 24-Core	2x AMD EPYC 7452 32-Core

[a]as measured via Graphcore's tool `gc-hosttraffictest` for measuring the host-device bandwidth on the IPU.

3.1 Model and Dataset

The benchmarked model was Scaled-YOLOv4-P5, adapted from the Graphcore example repository [5]. The model has approximately 70.8×10^6 parameters and requires approximately 167×10^9 multiply-add operations (MACs) per inference.[1] We used the COCO 2017 dataset's [12] 5,000-image validation set to verify accuracy using pre-trained weights [6], achieving $\text{mAP}_{50\text{–}95}$ results of 49.1

[1] Estimates using `torchinfo`.

and mAP_{50} of 68.7, consistent with the official benchmarks on both platforms [5].

The GPU port was gradually adapted to ensure maximum overlap to the original implementation on the IPU. This included replacing PopTorch dependency with PyTorch for the inference, explicit device management to shift all relevant tensors to the GPU, using `cuDNN` to find the most optimal convolution algorithms applicable for the model and elimination of redundant data movement between host and device during post-processing steps like non-maximum suppression (NMS).

3.2 Benchmarking Methodology

Two performance metrics were measured during inference (including on-device pre-processing and post-processing):

1. **Latency (ms)**: We measured full round-trip (end-to-end) latency. To ensure precision, we used device-specific timing primitives: `poptorch.PoplarExecutor.getLatency()` for the IPU and `torch.cuda.Event` for the GPU. This captures all host-to-device transfers, model execution, and device-to-host transfers. For the GPU, we used `torch.cuda.synchronize()` to account for CUDA's asynchronous nature.
2. **Throughput (FPS)**: We measured the total number of images processed per second. The total wall-clock time was measured using standard Python primitives (`time.time()`) on the host and divided by the total number of images, calculated after performing 100 warm-up iterations.

We varied three experimental parameters:

1. **Input Image Size**: We fixed the batch size to 1 and varied the input image sizes from 64 to 1280 pixels. This tests the hardware's scaling with spatial dimensions. The baseline batch size of 1 ensures the maximum possible image size can be tested on both hardware, along with providing a benchmark for real-life scenarios where batching is not feasible.
2. **Batch Size**: We fixed the image sizes and varied the batch sizes in powers of two from 1 to 64. Image sizes of 128, 224 and 320 pixels were used to provide a comprehensive overview. This tests the hardware's ability to exploit data parallelism.
3. **Floating-Point Precision**: We repeated all the experiments using both half-precision (FP16) and single-precision (FP32) floating point parameters to test the impact of numerical precision on memory footprint and compute throughput.

Lastly, to quantify the practical implications of each platform's development cycle, we measured the total time-to-result, defined as the sum of compilation and execution times. All results are reported as the average of 100 benchmark iterations.

4 Results

Once the accuracy of the model was verified on both platforms, we varied the image sizes and batch sizes. Preliminary tests revealed a failure on the IPU when compiling the model using a batch size of 2 using the model's default image size of 896 px due to the limited on-tile memory capacity of 624 KiB being exhausted. This memory limitation on the IPU constrains what image and batch sizes could be tested on both the hardware. Thus, the performance envelope of the IPU had to be determined before further tests could be performed, which is presented in Fig. 1.

When varying the input image sizes, we found the IPU being 3.99x faster than the GPU on the model's default input image size of 896 px, with a round-trip latency of 6.56 ms compared to the GPU's 26.17 ms when using a batch size of 1 at half precision. At batch size of 1, the IPU remained faster across all image sizes. The GPU maintained an almost constant latency until images of size 800 px, after which it trends upwards. At half precision, we observed a speedup ranging from 24.7x (image size = 128 px) till 3.60x (image size = 1088 px) on the IPU. The doubling of memory requirements when using a higher floating point precision level (FP32) resulted in the IPU successfully compiling only half of the tested image sizes along with a higher rate of increase in latency when increasing the input image size.

When varying the batch size on the IPU, smaller image sizes of 128 px, 224 px and 320 px were used due to its limited memory. The GPU maintained a consistent latency (at ~25 ms) from batch sizes of 1 to 8, confirming it is overhead-bound. The IPU's latency scales linearly with batch size (e.g. from 1.03 ms at batch 1 to 8.51 ms at batch 64 for 128 px images), confirming it is compute-bound. As shown in Fig. 4, the GPU's throughput scales almost linearly with batch size, whereas the IPU's scaling is shown to be weaker. Figure 3 shows the IPU always maintaining a lower latency than the GPU.

Doubling the precision leads to a cut in half for the IPU's maximum achievable batch size for most tests, as shown in Figs. 2, 3 and 4 along with a greater increase in latency on the IPU compared to the GPU, for which the latency and throughput remains the same despite using FP32 until the batch size of 8.

Finally, regarding development overhead, Fig. 5 shows the IPU taking 382.79 s (over 6 min) for a full benchmark run at 896 px, compared to the GPU's 15.56 s.

5 Discussion

As Fig. 1 illustrates, the NVIDIA A30's large HBM2 memory easily handled all tested batch sizes. In contrast, the Graphcore GC200 is severely constrained by its on-chip SRAM. At the model's native resolution, it can only compile for a batch size of 1; at 1120 px, it fails to compile entirely.

The performance divergence when varying image size stems directly from the architectural trade-offs. The IPU's 6.56 ms is the true compute time, as the model and graph are resident in fast SRAM. Thus, the IPU's latency, governed purely by computation, scales quadratically with the range of image sizes

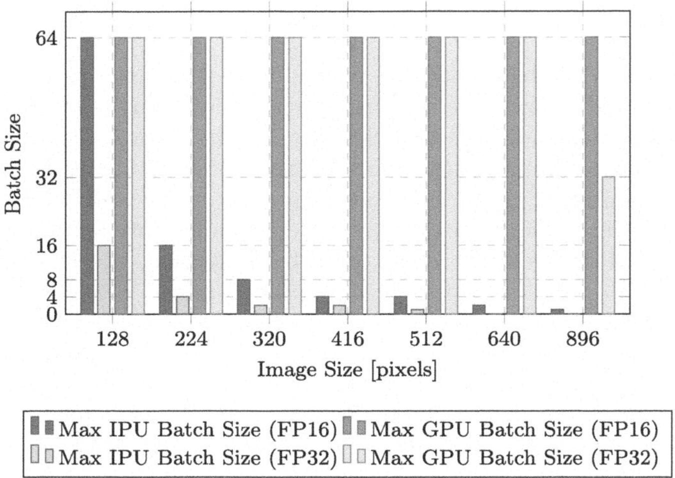

Fig. 1. Maximum achievable batch size for the Scaled-YOLOv4-P5 model on a set of pre-defined image sizes, contrasting the memory limitations of the Graphcore GC200 IPU to the Nvidia A30 GPU.

throughout, conforming to the theoretical relationship of quadratic scaling for increasing input image sizes on the YOLOv4 model [20]. The GPU's 26.17 ms is an overhead-bound time. Its SMs are starved for data at a batch size of 1, and the total time is dominated by kernel launches and off-chip memory fetching. For real-time, single-image applications, the MIMD/SRAM architecture is demonstrably superior.

Conversely, Figs. 3 and 4 demonstrate the GPU's SIMT advantage. Designed for massive data parallelism, it becomes fully saturated only at larger batch sizes, whereas the IPU's MIMD architecture and limited memory prevent it from scaling to the same batch sizes. In all the presented experiments, the IPU remains consistently bound more by compute than memory, unlike the GPU, which becomes compute-bound only at larger batch sizes, allowing throughput to scale linearly. This shows the GPU's advantage for applications where data can be batched and throughput is the primary metric to optimize.

Lastly, Fig. 5 illustrates how the NVIDIA A30 GPU's eager execution model and a Just-In-Time (JIT) execution approach is 24.6x faster compared to the IPU on the default image size. The Graphcore IPU, however, requires a complete, static computational graph before execution. The Poplar compiler performs an Ahead-of-Time (AOT) compilation to perform global optimizations and map the entire graph onto the IPU's 1,472 tiles, which incurs a substantial initial cost, although the resulting executable is cached for subsequent runs that share the same parameters.

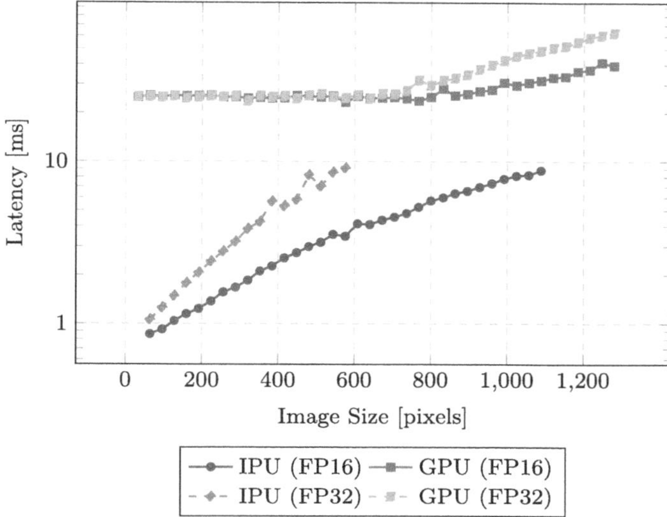

Fig. 2. Comparison of the effect of changing floating-point precision from FP16 to FP32 on the average single-batch inference latency as a function of the input image size. The throughput in this case is the inverse of latency given the fixed batch size of 1. The plot uses a log-linear scale.

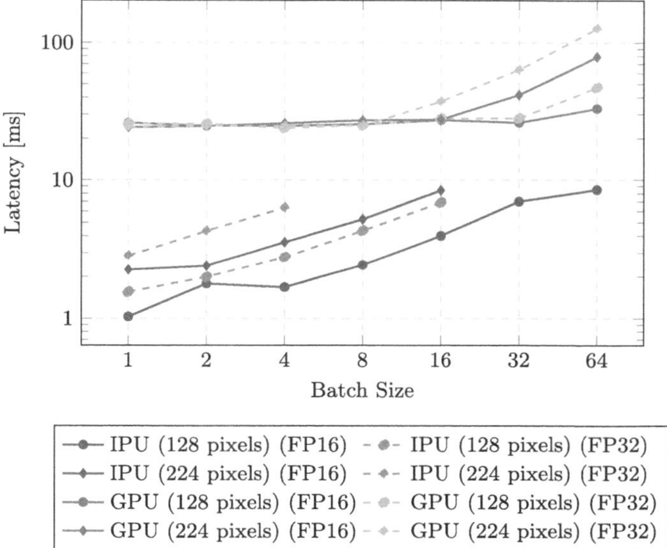

Fig. 3. Average inference latency as a function of batch size for both FP16 and FP32. This is shown using two image sizes to contrast not just the hardware, but also the effects of changing precision and image size on the same hardware. The plot uses a $\log_{10}$-$\log_2$ scale.

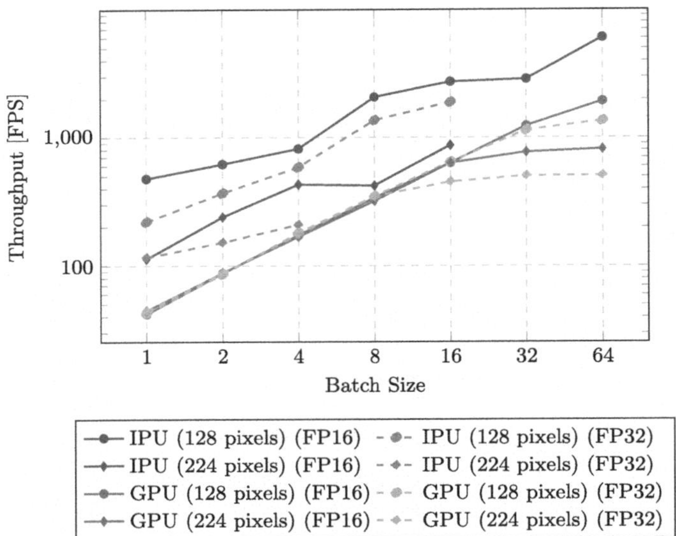

Fig. 4. Throughput vs. batch size across FP16 and FP32 precision for the image sizes of 128 and 224 pixels. The plot uses a $\log_{10}$-$\log_2$ scale.

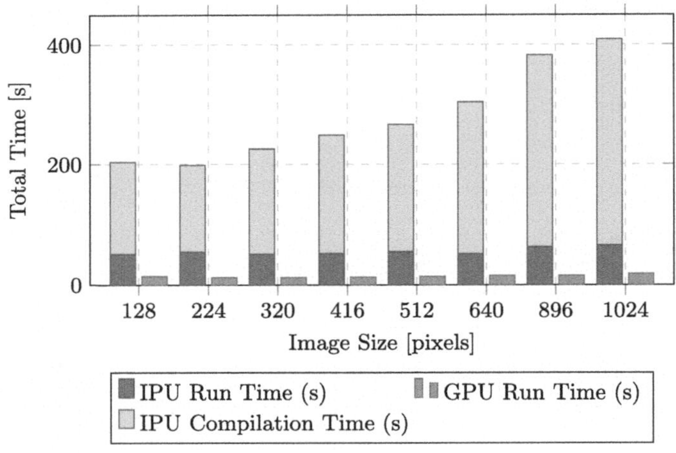

Fig. 5. Comparison of the total time-to-result on the IPU and the GPU. The IPU requires significant time to compile along with run-time overheads.

6 Conclusion

After verifying the feasibility of running an object detection model on the IPU, we show how the IPU excels at minimizing latency when using smaller image and batch sizes in contrast to the GPU, which excels at maximizing throughput with bigger image and batch sizes. The primary performance differentiator is in single-batch inference, where the IPU exhibits a significant advantage for real-time applications; at input image sizes of 896 px using FP16 precision, the IPU delivers a latency of 6.56 ms, an almost 4x speedup over the GPU's 26.17 ms. This low-latency advantage presents a trade-off, as the GPU's larger memory capacity of 24 GiB supports higher batch sizes (e.g., 32 vs. 8 at 320 px), enabling the GPU to achieve a greater peak throughput (426.17 FPS) compared to the IPU's memory-limited maximum (365.54 FPS) with $\sim$900MB of on-chip memory. Finally, due to the IPU's compilation cost, the total time to benchmark a new run was more than an order of magnitude higher than the GPU's. Nonetheless, for all independent metrics we benchmarked on the IPU where it was not constrained by its memory, it outperformed the GPU.

These findings suggest several avenues for future research. A comprehensive roofline analysis, for instance, could quantify performance bottlenecks, while energy efficiency could be assessed using metrics like FLOPS/Watt. Furthermore, this evaluation could be extended to multi-IPU configurations and newer models with diverse computational and memory demands. It would also be valuable to evaluate other models to assess whether the results can be generalized. The GC200 architecture also supports Streaming Memory, which was not utilized in this study. Future work could benchmark performance when streaming weights and activations from this larger memory pool to see if it provides a viable path for larger models or batches on the IPU. The results underscore the importance of aligning the architectural strengths of an accelerator with the specific requirements of the target application to maximize performance: latency (IPU) or throughput (GPU).

References

1. Arcelin, B.: Comparison of graphcore ipus and nvidia gpusfor cosmology applications (2021). https://arxiv.org/abs/2106.02465
2. Bohl, R.: Graphcore ipus: Accelerating argonne's ml/ai applications. https://extremecomputingtraining.anl.gov/wp-content/uploads/sites/96/2022/11/ATPESC-2022-Track-1-Talk-7-Bohl-Graphcore.pdf (2022). Presented at ATPESC 2022, Track 1 (Hardware Architectures). Accessed 2025-11-01
3. Gepner, P., Kocot, B., Paprzycki, M., Ganzha, M., Moroz, L., Olas, T.: Performance evaluation of parallel graphs algorithms utilizing graphcore IPU. Electronics **13**(11) (2024). https://doi.org/10.3390/electronics13112011. https://www.mdpi.com/2079-9292/13/11/2011
4. Graphcore: Performance results (2023). https://web.archive.org/web/20250813025206/https://www.graphcore.ai/performance-results. Accessed 2025-11-01
5. Graphcore: examples: vision/yolo_v4/pytorch (yolov4 pytorch example for graphcore ipus) (2025). https://github.com/graphcore/examples/tree/master/vision/yolo_v4/pytorch. GitHub repository directory; Accessed 2025-11-01
6. Graphcore: Yolov4 p5 reference weights (2025). https://gc-demo-resources.s3.us-west-1.amazonaws.com/yolov4_p5_reference_weights.tar.gz. archive of YOLOv4 P5 model weights; Accessed 2025-11-01

7. Graphcore Limited: Benchmarks. Tech. rep., Graphcore Limited (2020). https://www.graphcore.ai/hubfs/assets/pdf/GC_Benchmarks_Oct_20.pdf?hsLang=en, pDF available at https://www.graphcore.ai/hubfs/assets/pdf/GC_Benchmarks_Oct_20.pdf?hsLang=en
8. He, K., Zhang, X., Ren, S., Sun, J.: Deep residual learning for image recognition (2015). https://arxiv.org/abs/1512.03385
9. Hooker, S.: The hardware lottery (2020). https://arxiv.org/abs/2009.06489
10. Jia, Z., Tillman, B., Maggioni, M., Scarpazza, D.P.: Dissecting the graphcore ipu architecture via microbenchmarking (2019). https://arxiv.org/abs/1912.03413
11. John, C.M., Nassyr, S., Penke, C., Herten, A.: Performance and power: Systematic evaluation of ai workloads on accelerators with caraml. In: SC24-W: Workshops of the International Conference for High Performance Computing, Networking, Storage and Analysis, pp. 1164–1176. IEEE (2024). https://doi.org/10.1109/scw63240.2024.00158
12. Lin, T.Y., et al.: Microsoft coco: Common objects in context (2015). https://arxiv.org/abs/1405.0312
13. Liu, S., Qi, L., Qin, H., Shi, J., Jia, J.: Path aggregation network for instance segmentation (2018). https://arxiv.org/abs/1803.01534
14. Shekofteh, S.K., Alles, C., Fröning, H.: Reducing memory requirements for the ipu using butterfly factorizations. In: Proceedings of the SC '23 Workshops of the International Conference on High Performance Computing, Network, Storage, and Analysis, pp. 1255–1263. SC-W '23, Association for Computing Machinery, New York, NY, USA (2023). https://doi.org/10.1145/3624062.3624196
15. Shekofteh, S.K., Alles, C., Kochendörfer, N., Fröning, H.: On performance analysis of graphcore ipus: Analyzing squared and skewed matrix multiplication (2023). https://arxiv.org/abs/2310.00256
16. Shekofteh, S.K., Bogacz, D., Alles, C., Fröning, H.: Butterfly factorization for vision transformers on multi-IPU systems. Parallel Comput. **127**, 103165 (2026). https://doi.org/10.1016/j.parco.2025.103165
17. Sumeet, N., Rawat, K., Nambiar, M.: Performance evaluation of graphcore ipu-m2000 accelerator for text detection application. In: Companion of the 2022 ACM/SPEC International Conference on Performance Engineering. pp. 145–152. ICPE '22, Association for Computing Machinery, New York, NY, USA (2022). https://doi.org/10.1145/3491204.3527469
18. Tan, M., Le, Q.V.: Efficientnet: Rethinking model scaling for convolutional neural networks (2020). https://arxiv.org/abs/1905.11946
19. Valiant, L.G.: A bridging model for parallel computation. Commun. ACM **33**, 103–111 (1990). https://api.semanticscholar.org/CorpusID:15655597
20. Wang, C.Y., Bochkovskiy, A., Liao, H.Y.M.: Scaled-yolov4: Scaling cross stage partial network (2021). https://arxiv.org/abs/2011.08036
21. Wang, C.Y., et al.: Cspnet: A new backbone that can enhance learning capability of CNN (2019). https://arxiv.org/abs/1911.11929
22. Xie, S., Girshick, R., Dollár, P., Tu, Z., He, K.: Aggregated residual transformations for deep neural networks (2017).https://arxiv.org/abs/1611.05431

FPGA-Based Anomaly Detection for Medical IoT: Trade-Offs Between Hls4ml and FINN

Sayanti Pal(✉), Maximilian Krentzien, Florian Rokohl, and Marc Reichenbach

Institute for Applied Microelectronics and Data Technology, University of Rostock, Rostock, Germany
{sayanti.pal,maximilian.krentzien,florian.rokohl, marc.reichenbach}@uni-rostock.de

Abstract. The growing demand for secure healthcare services in medical IoT networks has raised the need for robust anomaly detection systems, with autoencoder (AE) as a compelling solution. Although mapping of neural networks on FPGAs is well-known, selecting a suitable High-Level Synthesis (HLS) framework for a particular use case remains a challenge. This work explores the trade-offs between two well-known, open-source HLS frameworks: hls4ml and FINN by evaluating non-functional metrics for deploying AEs. Our results highlight that hls4ml achieves roughly 10× lower latency and 3× higher throughput as compared to FINN. On the other hand, FINN provides a comprehensive, automated end-to-end solution for deploying quantized networks on FPGA with up to 80% fewer lookup tables (LUTs) and flip-flops (FFs), but with limited control over the design process. Indicating that FINN is suitable for low-power deployments where faster time-to-deployment is required, whereas hls4ml excels in latency-critical, high bandwidth applications, our study sheds light on the choice of HLS tools.

Keywords: Anomaly Detection · Edge Deployment · FPGA · FINN · hls4ml · High-Level Synthesis · Low-Latency Inference · Quantization · Hardware-Software Co-design

1 Introduction

The widespread use of medical Internet of Things (IoT) caused groundbreaking advancements in remote health monitoring, patient care, and streamlining hospital workflows [19], while introducing cybersecurity threats such as data tampering and unauthorized access. This can lead to undetected anomalies like intrusion deviations and might cause a life-threatening situation or compromise sensitive medical data [14]. These risks intensify in 5G-enabled medical IoT as it handles massive connectivity with a strict latency budget. As a solution to this, real-time, machine learning (ML)-powered edge anomaly detection has emerged as a crucial component [19].

M. Kowalczyk and C. Monière (Eds.): DASIP 2026, LNCS 16512, pp. 119–131, 2026.
https://doi.org/10.1007/978-3-032-23871-9_10

However, implementing these ML models in real-time, resource-constrained edge environments remains a challenge. Field-Programmable Gate Arrays (FPGAs) are well-suited for this task due to their deterministic low-latency performance, reconfigurability, and parallel processing capabilities. When compared to CPU and GPU implementations for comparable latency and energy efficiency, FPGA-accelerated inference is preferable due to its specialized dataflow architecture. For example, FPGA for binary convolutional neural networks (CNNs), can achieve 8.3× higher performance and 75× greater energy efficiency compared to a Titan X GPU [15]. For a swin transformer model, FPGA accelerator achieves significantly higher energy efficiency, with up to 20.45× over the AMD Ryzen 5700X CPU and 5.05× over the Nvidia GeForce RTX 2080 Ti GPU [16], revealing FPGA's suitability for edge-inference. Deploying neural networks on FPGAs is a challenging and time-consuming task that requires manual conversion of models to hardware description language (HDL), demanding knowledge of both hardware and software design [18]. Converting sequential algorithms into parallelized structures while balancing hardware resource constraints is both labour-intensive and prone to design inefficiencies and integration errors. This can significantly delay the time-to-market and deployment for cutting-edge AI solutions. To address this, High-Level Synthesis (HLS) frameworks, a well-known approach in other domains, have become important by automating the conversion process, preserving accuracy, and enabling faster development cycles [18]. HLS frameworks help to implement neural networks on FPGA using familiar high-level programming languages like C/C++ with automated conversions from PyTorch or TensorFlow to optimized HDL, utilizing important network components like convolutional layers, activation functions, and batch normalization.

Two well-known, open-source HLS frameworks, hls4ml and FINN, are chosen in this work due to their ability to generate custom FPGA architectures with optimized quantization support and rapid prototyping. The requirements for high energy efficiency, optimized resource utilization, and low-latency make edge anomaly detection a prime use case for evaluating the effectiveness of these HLS frameworks. Rather than focusing on algorithmic evaluation, this study offers two significant contributions: (i) an analysis of fundamentally distinct hardware architectures generated by hls4ml and FINN under the same workload and matched hardware constraints; and (ii) a controlled, reproducible head-to-head benchmark of the two tools' non-functional metrics, providing clear guidance on 'when-to-use-which' for edge deployments.

2 Related Work

This overview covers advancements in three interconnected fields: anomaly detection in medical IoT systems with state-of-the-art, FPGA-based ML deployment, and the significance of HLS tools like hls4ml and FINN.

Recent studies highlight that the medical IoT's growth has enhanced healthcare [19] while introducing security risks. ML can improve security and dependability [5], as demonstrated by ensemble AI models for anomaly detection [7], which improve irregularity identification using the CICIoMT2024 dataset. As 5G-enabled medical IoT systems need efficient edge-inference of these ML models on FPGA, an effective deployment approach is essential. HLS frameworks have emerged as a promising solution to that, outperforming manual HDL conversion and [18]. With advancements in compilation technologies and platform-based modelling, HLS frameworks have successfully been applied to deploy convolutional AEs [9], and variational AEs [20]. With the significance of HLS frameworks, recent research has focused on improving and investigating specialized HLS frameworks, FINN and hls4ml, for edge deployments. Although direct comparisons between these two tools are limited, each framework has been thoroughly examined on its own. FINN's register-transfer level (RTL) implementation and its feasibility in deploying neural networks with streaming architecture are emphasized in [2]. Performance and resource usage across different FPGA platforms with quantized networks [10], and low-latency recurrent neural network inference using FINN [1] have been studied thoroughly. Meanwhile, hls4ml has been used for implementing low-latency, low-power CNNs on FPGAs [1] and mapping them onto interconnected FPGA devices [17]. These studies show how effective FINN and hls4ml are to deploy complicated models on FPGAs while maintaining high performance and low latency, making them appropriate for real-time applications. Alongside FINN and hls4ml, Vitis AI has also been studied for its suitability for FPGA-based Deep Neural Network (DNN) acceleration. [12] highlights that FINN significantly outperforms the Vitis-AI-based accelerator, attaining 3.3× better power efficiency and 3.0× better throughput. [11] presents that FINN and hls4ml focus on low-precision quantization, hardware-software co-design, and shortened development times for ML workloads on FPGAs, in contrast to general-purpose overlay-based DNN deployment toolchains like Vitis and TVM VTA. Also, further optimization like quantization-aware training (QAT) with FINN for quick prototyping is explored in [8].

This background sets the foundation for our work that bridges the gap in the literature, providing a non-functional evaluation of FINN and hls4ml. Furthermore, multiplier-less dataflow-architectures have been explored in linear computation-coding (LCC) based approaches [3]. This work extends those architectural explorations by performing an application-oriented evaluation.

3 Methodology and Implementations

This section presents our pipeline for comparing implementations of hls4ml and FINN which consists of a) development and QAT of an ML-model, b) functional evaluation of the model and c) optimized implementation on FPGA. In this work, we are accelerating the inference by improving the efficiency of MAC execution, as shown in Fig. 1.

3.1 Model Architecture and Data Set

AEs cover multiple fields of applications as shown by [6]. Among other neural network architectures, AEs have emerged as a promising option in anomaly

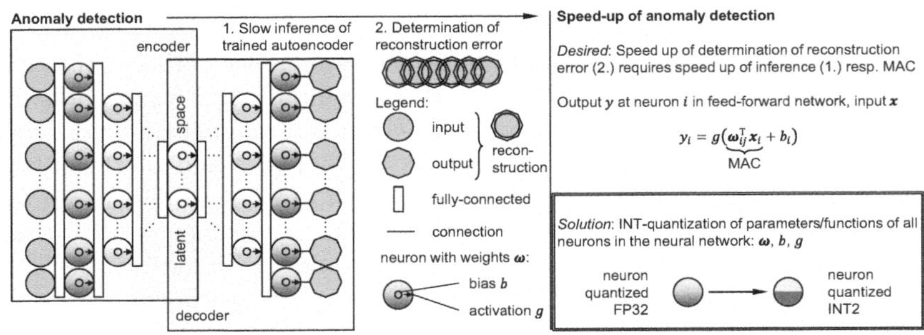

Fig. 1. Speed-up of anomaly detection using autoencoder.

detection [21]. For this work, an AE model has been selected that consists of consecutive fully-connected dense layers presented in Table 1.

Table 1. Neural network architecture of the autoencoder

Layer Domain	Layer	#Neurons	Activation
Encoder	1, 2, 3	99, 54, 26	*ReLU, ReLU, ReLU*
Latent Space	4	26	*ReLU*
Decoder	5, 6, 7	26, 54, 99	*ReLU, ReLU, tanh*

Considering bias, the model yields 25,037 trainable parameters, making it sufficiently expressive and compact, suitable for low-latency, power-efficient inference in typical edge settings. Although *ReLU* is optimal for hardware implementation, functional-wise a signed activation function, such as *tanh*, is well-suited for reconstructing negative inputs. The model is trained with publicly available *5GC PFCP Intrusion Detection Dataset* [4], addressing DoS (denial of service) attacks that are relevant for the medical domain.[1] Within the dataset, the subset of *Balanced TCP-IP Layer/20-sec-CSV* is utilized, and the original train-test split is kept as already existing.

3.2 Training, Quantization, and Functional Evaluation

The data preparation process leads to 99 features. Then, an additional split of the original training data into training and validation data has been conducted with a ratio of 17/3. After that, QAT has been utilized via QKeras/TensorFlow and Brevitas/Pytorch for hls4ml and FINN, respectively. The model is trained on normal (non-malicious) data only; afterwards, the test data of normal and malicious samples are reconstructed by the trained AE. The reconstruction error

[1] https://www.radware.com/blog/ddos-protection/5-ways-ddos-attacks-disrupt-healthcare-services/.

for fixed-point models is derived from the mean squared error metrics by comparing the reconstructed output as a fixed-point value against the input of the AE that exists as a floating-point value, as shown in Fig. 1. With the results of the reconstruction error distribution of normal test data, the threshold is defined as follows: the sum of the average value and the standard deviation (referred to as "1 sigma strategy" in Fig. 2) of normal test data reconstruction errors. In effect, a classifier is defined: values greater than or equal to this threshold are considered to be malicious and below to be normal, as shown in Fig. 2.

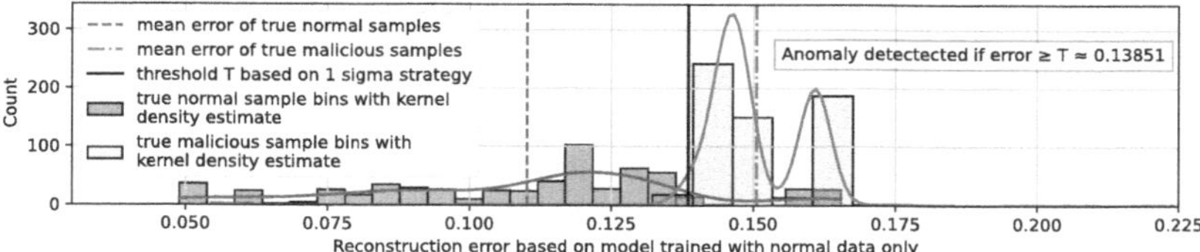

Fig. 2. Reconstruction error distribution of test data.

Based on an evaluation with Keras and hls4ml, the normal test data and the malicious test data with the trained AE, the F_1 scores are presented in Table 2. The results show that the model is suitable for detecting anomalies.

Table 2. Functional evaluation in Keras and hls4ml

Precision of AE	Abbreviation	F_1-score [%]	Framework
32-bit floating point	FP32	94.20	Keras
8-bit fixed point	INT8	95.19	QKeras/hls4ml
4-bit fixed point	INT4	42.95	QKeras/hls4ml

Note that models with precision $<$ 8 bit are not optimized for functional performance, but they are included to evaluate the hardware implications with quantization. We focus on non-functional metrics, leaving functional enhancements like alternative quantization strategies for future work.

3.3 HLS Workflow and Optimizations

FINN and hls4ml generate optimized hardware accelerators from PyTorch and TensorFlow trained models, respectively. Figure 3 illustrates the hardware-software co-design flow for building FPGA-based accelerators, which performs optimization of the trained model through HLS, combining C++ simulation, logic synthesis, hardware-specific optimizations like weight and activation quantization, batch normalization folding, and memory access optimization, followed by FPGA deployment. To further optimize performance and resource utilization, several strategies inside the hls4ml and FINN pipeline have been used.

3.3.1 Hls4ml Optimizations

hls4ml uses several built-in optimization strategies to convert the model into a high-performance architecture. It allows setting layer-specific reuse factors and implementation strategy, such as *Latency* and *Resource* [13]. The reuse factor controls the degree of loop unrolling in generated HLS code by determining the number of times a hardware multiplier or MAC unit is reused. In FP32 model, relatively high reuse factors have been selected (in the range of 50–100), while in quantized models, they are much lower (around 10–30) to exploit parallelism and ensure efficient mapping without exceeding resource budgets. The reuse factors rise gradually from the layer in the latent space outwards in both encoder and decoder directions to optimize resource usage, allocating more resources to layers handling larger volumes of data. This way, hls4ml offers high flexibility to select reuse factors for adapting a wide range of model sizes and quantization schemes. As the *Latency* strategy aims to increase parallelism, higher throughput and reduced latency have been observed for AE models. Still, it comes at the cost of higher resource usage. Therefore, *Resource* strategy has been set for our dense AE variants to strike a balance between resource and latency.

The architecture here is built on the foundation of inter-layer temporal pipelining and intra-layer spatial parallelism. This allows multiple neuron computations to be executed concurrently within each dense layer, utilizing dedicated DSP and LUT as determined by the reuse factor, which in turn enables parallel MAC operations and reduces latency.

Additionally, hls4ml's dataflow scheduling strategy connects layers through streaming FIFOs and creates a deep pipeline as shown in Fig. 5(a), allowing for simultaneous processing of multiple input samples and achieving continuous throughput of one output per cycle in steady state. This approach outperforms resource-folded designs in terms of throughput. The default hls4ml flow enables aggressive loop-unrolling with a wide, fully parallel input/output interface,

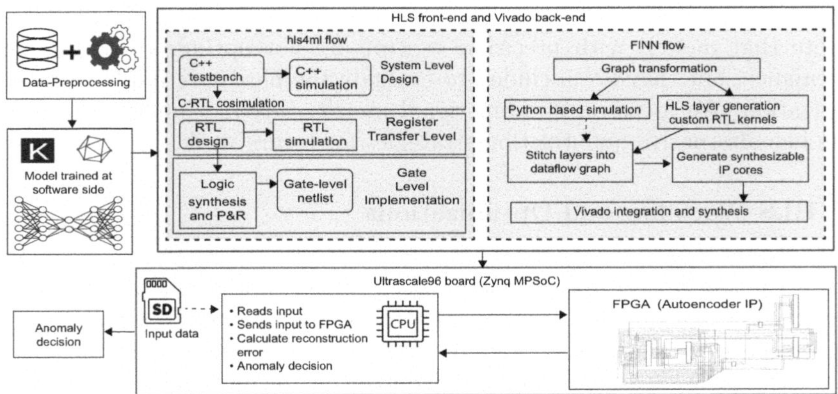

Fig. 3. Hardware-software co-design for anomaly detection using HLS frameworks.

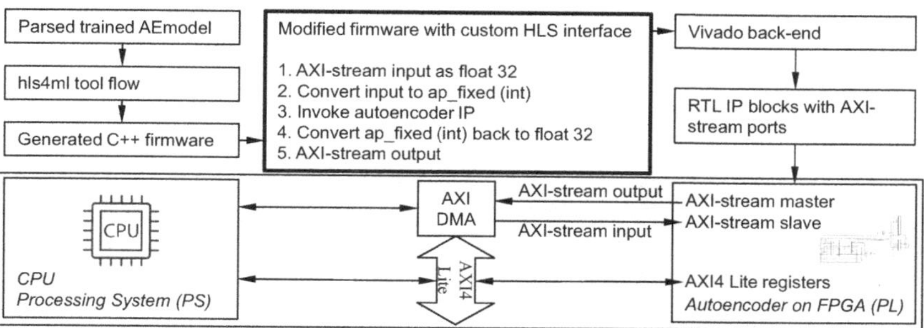

Fig. 4. hls4ml workflow with AXI stream approach.

resulting in over-utilization of resources. For our dense AE, inputs/outputs are streamed via FIFOs with partial loop-unrolling, and the intermediate firmware has been modified for our AXI width and input data width constraints. Our custom wrapper converts AXI4-Stream input to fixed-point and invokes the AE kernel for execution. After processing, it converts the output back to floating-point and streams it out using AXI DMA (Direct Memory Access) IP. Additionally, control signals are made accessible through a compact AXI4-Lite register set. Figure 4 shows the hls4ml flow leveraging AXI stream protocol, which streams inputs directly to the first layer, increasing cycles per layer.

3.3.2 FINN Optimizations

FINN's end-to-end workflow generates a streaming dataflow architecture in which each neural network layer is implemented as an independent hardware operator, interconnected by AXI-Stream FIFOs, as shown in Fig. 5(b). To generate a streamlined dataflow graph, our model has been lowered through a sequence of intermediate-representation graph transformations that fuse each dense (GEMM), bias, and activation block into a single quantized Matrix Vector Activation Unit (MVAU) node. MVAU is the core computational block for linear layers, which efficiently performs matrix-vector multiplication (MVM), minimizing resource utilization and power consumption [2]. Spatial parallelism in MVAUs is controlled by tuning the number of processing elements (PEs) and Single Instruction Multiple Data (SIMD) lanes per layer. While wider SIMD lanes help to accelerate the layer-multiplications, computations are parallelised across multiple neurons by increasing the number of PEs. FINN allows setting a specific throughput target or frames per second (fps), and based on that, folding parameters-PE and SIMD can be selected. Two distinct fps objectives are targeted as representative design points in our experiments to implement the models: design A and design B represent 0.5Mfps and 2Mfps, respectively.

Table 3 shows the per-layer folding and corresponding cycle counts for both design configurations, which result in aggressive parallelism. Parallelism is further enhanced in design B, where the cycle count for critical layers is reduced

Table 3. Per-layer folding and cycle counts across two FINN configurations. Layer cycles are denoted as LC. Subscripts A and B denote design A and B.

Layer	PE_A	$SIMD_A$	LC_A	PE_B	$SIMD_B$	LC_B
MVAU_hls_0	3	9	198	18	11	27
MVAU_hls_1	1	9	162	3	18	27
MVAU_hls_2	1	9	81	1	27	27
MVAU_hls_3	1	9	81	1	27	27
MVAU_hls_4	1	9	162	6	9	27
MVAU_rtl_0	3	9	198	9	18	33

from 198 to 27, significantly reducing the overall critical path. As a result, latency reduces from 8.8μs to 1.6μs in design B, and also, throughput improves 6× as compared to design A. When implemented at 100MHz, high theoretical throughput is reported based on the target fps and folding parameters. As the fps targets are too high for our small fully-connected AE, it is achieved at a cost of higher resource usage. The FIFOs manage backpressure through standard AXI-Stream handshaking, enabling the MVAUs to operate in a continuous, pipelined dataflow. Leveraging FINN's deployment package, an end-to-end inference runtime has been generated that automates the stitched IP generation and FPGA acceleration.

While both frameworks aim for optimized hardware implementation, they reflect their unique architectural approaches employing distinct strategies. FINN focuses on dataflow-oriented transformations and hardware-specific primitives for quantized networks, whereas hls4ml relies on reuse-based parallelism with a single-core HLS approach.

3.3.3 Synthesis Strategies and Critical Path Handling

To ensure a fair comparison, both FINN and hls4ml-based designs are synthesized and implemented in Vivado 2024.2 with 100MHz clock frequency applied to the programmable logic (PL) domain. The post-route timing reports have revealed that, despite sharing the same clock constraint, the tools achieved different maximum frequencies, which is attributable to the fundamentally distinct hardware architectures and critical path structure generated by them.

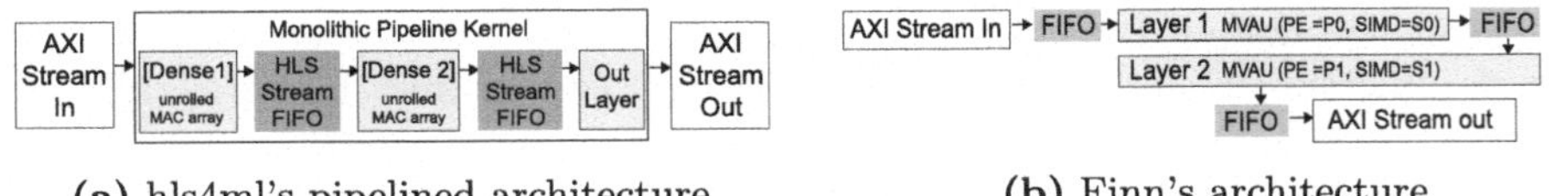

(a) hls4ml's pipelined architecture **(b)** Finn's architecture

Fig. 5. generated architecture by both tools.

hls4ml generates a heavily pipelined compute kernel, with dense arithmetic and control logic encapsulated within a single large module. To manage timing, it relies on HLS-level directives such as pipelining, loop unrolling, and array partitioning, which insert pipeline registers to shorten logic depth. Despite resulting in very low latency, the residual critical paths are often dominated by routing delays due to the wide datapaths and high fan-out inside dense layers, leading to a relatively lower maximum frequency compared to FINN.

In contrast, FINN adopts a layer-wise dataflow architecture with each layer becoming a self-contained hardware block connected through AXI-Stream FIFO. This approach allows long combinational paths to be localized within single MVAU operators, resulting in critical paths characterized by short LUT/CARRY reduction chains with routing-dominated delays. In order to meet timing requirements, FINN relies primarily on physical optimizations, including placement, buffering, and replication, achieving high frequencies for low-precision designs due to its modular structure.

4 Results and Discussions

The AE variants are implemented on the Avnet Ultra96 (Xilinx Zynq UltraScale+ MPSoC), using hls4ml and FINN. This low-power MPSoC tightly integrates ARM Cortex-A53 PS and PL via AXI-DMA, making it a suitable option for implementing a heterogeneous, CPU-FPGA-based anomaly detection system, as described in Fig. 3. Evaluations have been done with same AXI interface width, and the same clock frequency, ensuring a fair comparison. Noatbly, FINN supports fixed-point models only; hls4ml accepts floating-point models but lowers them to fixed-point for HLS, 16 bit in our case.

Table 4. Comparing design characteristics, achieved $f_{\max}$, latency, throughput at $f_{\max}$.

Aspect/Precision	hls4ml			FINN		
	INT2	INT4	INT8	INT2	INT4	INT8
$f_{\max}$ (MHz)	128	116	116	240	227	103
Throughput (GOPS)	150.3	80.6	9.6	14.2	10.3	1.12
Latency (μs)	0.813	1.517	4.60	18.8	20.4	66.3
Design structure	Monolithic HLS kernel			Layer-wise dataflow		
Critical path	Dense-layer routing			MVAU MAC chain		

With 100MHz clock constraint, FINN's modular design achieves higher $f_{\max}$ for INT4 and INT2 models than hls4ml's monolithic kernel; a summary of the maximum frequencies achieved by both tools is illustrated in Table 4. But, FINN's per-inference cycle count is much larger, so at $f_{\max}$, it delivers significantly higher latency and lower throughput compared to hls4ml. FINN's higher clock does not offset its heavier cycle budget; hls4ml's pipeline with small initiation interval remains faster and substantially more throughput-dense even when each flow is measured at its own maximum clock. As these differences arise from internal design philosophies rather than unequal synthesis conditions, comparing non-functional metrics becomes meaningful at 100MHz operating point.

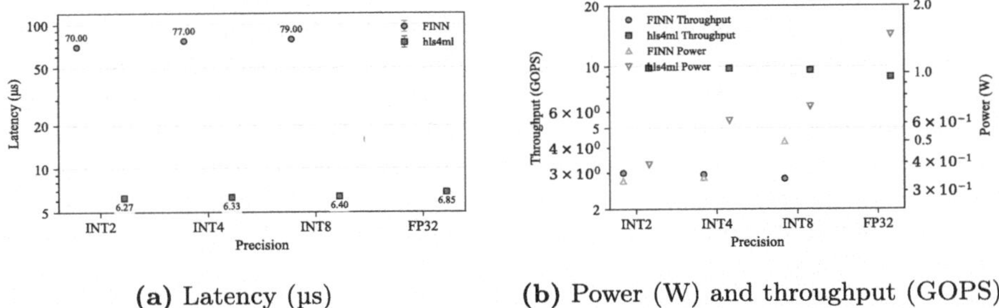

(a) Latency (µs) **(b)** Power (W) and throughput (GOPS)

Fig. 6. hls4ml vs. FINN: (a) latency comparison; (b) power and throughput comparison.

As Fig. 6(a) shows, hls4ml is 10–12× faster than FINN when comparing latency. The INT8 model achieves 6.4µs with hls4ml and 79µs with FINN, making hls4ml a better choice for real-time applications requiring ultra-low latency.

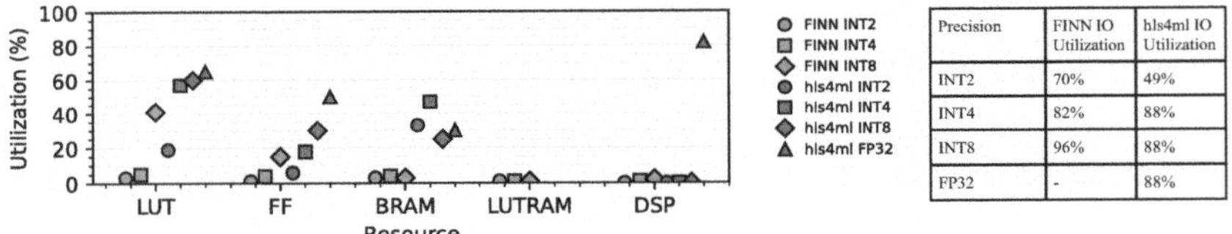

Precision	FINN IO Utilization	hls4ml IO Utilization
INT2	70%	49%
INT4	82%	88%
INT8	96%	88%
FP32	-	88%

Fig. 7. Core resource usage with IO% (AE only IO% assuming off-chip integration).

Figure 6(b) demonstrates that hls4ml outperforms FINN in terms of computational throughput, showing that the INT8 model achieves 9.8GOPS and 2.81GOPS for hls4ml and FINN, respectively. In terms of throughput per unit of power consumption (GOPS/W), hls4ml achieves 2–3× higher energy-efficiency than FINN. FINN consumes more power at higher $f_{\max}$, but per MHz power remains comparable for both tools. The trade-off between FINN and hls4ml in terms of resource utilization is shown in Fig. 7. FINN is more resource-efficient, consuming fewer core resources (LUTs, FFs, and BRAMs), compared to 57% LUTs and 18% FFs for hls4ml. Additionally, LUT and FF utilization are reduced for the models INT8, INT4, and INT2 compared to the FP32 variant with hls4ml. While our study primarily focuses on the AE described in Table 1, we have additionally evaluated a shallower variant of the primary AE, derived by removing layer 3, 5, and with a sigmoid activation function in the decoder output. On the same FPGA platform under the same clock constraint, the model shows the same trends with hls4ml achieving 10× faster latency and 3× higher throughput than FINN. While both tools show a comparable power consumption for this model, resource utilization follows the same trend as the primary AE, as shown in Table 5.

Table 5. Shallow autoencoder: throughput (Thr.) and core resources with IO.

Precision	FINN				hls4ml			
	Thr. (GOPS)	LUT	FF	IO	Thr. (GOPS)	LUT	FF	IO
INT2	3.5	2%	2%	70%	10.2	7%	3%	45%
INT4	3.5	3%	2%	88%	9.8	8%	4%	48%
INT8	2.9	17%	6%	94%	9.5	11%	4%	49%
FP32	–	–	–	–	9.0	11%	4%	49%

Figure 7 and Table 5 show Vivado's IO% with the accelerator as the top module, which represents the top-level ports with wrapper/interface complexity but not the actual IO usage on our MPSoC. With identical AXI payload width and fixed pin budget, hls4ml core uses fewer top-level streams and less IO% than the FINN stitched core. hls4ml's consolidated wrapper is easier to export off-chip, whereas FINN's wider wrapper limits exportability with a complex interface to integrate as a stand-alone IP.

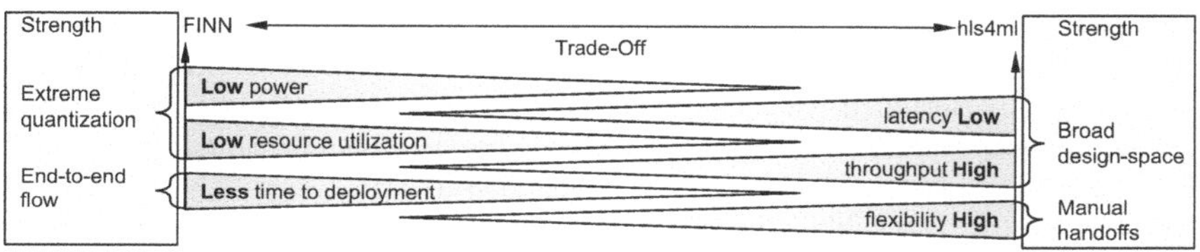

Fig. 8. Trade-Offs between FINN and hls4ml based on their strengths.

These results demonstrate clear trade-offs between the frameworks, as highlighted in Fig. 8.

5 Conclusion and Future Work

A comparative analysis of FINN and hls4ml for accelerating the inference of AE for FPGA-based anomaly detection has been presented in this work. hls4ml doesn't provide an end-to-end flow, leaving the integration of the IO interface explicitly up to the developer. In contrast, FINN offers a complete end-to-end flow with limited flexibility over the design process, providing less time-to-deployment. hls4ml demonstrates significantly faster latency and higher throughput, while FINN offers a resource-efficient solution, making the selection of a framework application-specific. This study lays the groundwork for future research into exploring design space by addressing threshold determination and hls4ml-internal quantization strategies. Also, our future work will focus on comparing high-throughput dataflow accelerators using LCC and HLS methodologies.

Acknowledgement. This work was funded by the German Federal Office for Information Security (BSI) under grant number 01MO23015D and partly funded by the Deutsche Forschungsgemeinschaft (DFG, German Research Foundation) – 459154262.

References

1. Aarrestad, T., Loncar, V., Ghielmetti, N., Pierini, M., Summers, S., Ngadiuba, J., et al.: Fast convolutional neural networks on FPGAS with HLS4ML. Mach. Learn. Sci. Technol. **2**(4), 045015 (2021). https://doi.org/10.1088/2632-2153/ac0ea1
2. Alam, S.A., Gregg, D., Gambardella, G., Preusser, T., Blott, M.: On the RTL implementation of FINN matrix vector compute unit (2022). https://arxiv.org/abs/2201.11409
3. Almeeva, L., Lehnert, A., Müller, R., Reichenbach, M.: High-performance, area-efficient and predictable matrix multiplication for ASIC. In: 2025 IEEE 36th International Conference on Application-specific Systems, Architectures and Processors (ASAP), pp. 182–183 (2025). https://doi.org/10.1109/ASAP65064.2025.00042
4. Amponis, G., et al.: 5G core PFCP intrusion detection dataset. In: 12th International Conference on Modern Circuits and Systems Technologies (MOCAST), pp. 1–4. IEEE (2023). https://doi.org/10.1109/MOCAST57943.2023.10176693
5. Aversano, L., Bernardi, M., Cimitile, M., Montano, D., Pecori, R., Veltri, L.: Anomaly detection of medical Iot traffic using machine learning, pp. 173–182 (2023). https://doi.org/10.5220/0012132000003541
6. Berahmand, K., Daneshfar, F., Salehi, E.S., Li, Y., Xu, Y.: Autoencoders and their applications in machine learning: a survey. Artifi. Intell. Rev. **57** (2024). https://api.semanticscholar.org/CorpusID:267466773
7. Chandekar, P., Mehta, M., Chandan, S.: Enhanced anomaly detection in IOMT networks using ensemble ai models on the ciciomt2024 dataset (2025). https://arxiv.org/abs/2502.11854
8. Chawda, D., Senouci, B.: Fast prototyping of quantized neural networks on an FPGA edge computing device with Brevitas and FINN. In: 2024 Fifteenth International Conference on Ubiquitous and Future Networks (ICUFN), pp. 238–240 (2024). https://doi.org/10.1109/ICUFN61752.2024.10625618
9. Dao-Xuan, S., Nghiem-Tuan, K., Truong-Dai, D., Ong-Tung, A., Hoang-Phuong, C., Nguyen-Duc, M.: Implementing convolutional auto encoder on FPGA using high level synthesis. In: Proc. 10th International Conference on Communications and Electronics (ICCE) (2024). https://doi.org/10.1109/ICCE62051.2024.10634692

10. Ducasse, Q., Cotret, P., Lagadec, L., Stewart, R.: Benchmarking quantized neural networks on FPGAS with FINN. CoRR abs/2102.01341 (2021). https://arxiv.org/abs/2102.01341
11. Ghiglione, M., Serra, V., Raoofy, A., Dax, G., Trinitis, C., Werner, M., et al.: Survey of frameworks for inference of neural networks in space data systems. In: DASIA 2022: Data Systems in Aerospace. Eurospace (2022). https://api.semanticscholar.org/CorpusID:259504103
12. Hamanaka, F., Odan, T., Kise, K., Chu, T.V.: An exploration of state-of-the-art automation frameworks for FPGA-based DNN acceleration. IEEE Access **11**, 5701–5713 (2023). https://doi.org/10.1109/ACCESS.2023.3236974
13. Khoda, E.E., Rankin, D., Teixeira de Lima, R., Harris, P., Hauck, S., Hsu, S., et al.: Ultra-low latency recurrent neural network inference on FPGAs for physics applications with hls4ml. Mach. Learn. Sci. Technol. **4**(2), 025004 (2023). https://doi.org/10.1088/2632-2153/acc0d7
14. Kruse, C., Frederick, B., Jacobson, T., Monticone, D.: Cybersecurity in healthcare: a systematic review of modern threats and trends. Technol. Health Care **25**, 1–10 (2016). https://doi.org/10.3233/THC-161263
15. Li, Y., Liu, Z., Xu, K., Yu, H., Ren, F.: A GPU-outperforming FPGA accelerator architecture for binary convolutional neural networks. J. Emerg. Technol. Comput. Syst. **14**(2) (2018). https://doi.org/10.1145/3154839
16. Liu, Z., Yin, P., Ren, Z.: An efficient FPGA-based accelerator for Swin transformer (2023). https://arxiv.org/abs/2308.13922
17. Mageiropoulos, E., Chrysos, N., Dimou, N., Katevenis, M.G.H.: Using hls4ml to map convolutional neural networks on interconnected FPGA devices. In: 2021 IEEE 29th Annual International Symposium on Field-Programmable Custom Computing Machines (FCCM), pp. 277–277 (2021). https://api.semanticscholar.org/CorpusID:235308027
18. Millón, R., Frati, E., Rucci, E.: A comparative study between HLS and HDL on soc for image processing applications. Elektron **4**(2), 100–106 (2020). https://doi.org/10.37537/rev.elektron.4.2.117.2020
19. Osama, M., et al.: Internet of medical things and healthcare 4.0: trends, requirements, challenges, and research directions. Sensors **23**, 7435 (2023). https://doi.org/10.3390/s23177435
20. Que, Z., Zhang, M., Fan, H., Li, H., Guo, C., Luk, W.: Low latency variational autoencoder on FPGAS. IEEE J. Emerging Selected Top. Circuits Syst. **14**(2), 323–333 (2024). https://doi.org/10.1109/JETCAS.2024.3389660
21. Zamanzadeh Darban, Z., Webb, G.I., Pan, S., Aggarwal, C., Salehi, M.: Deep learning for time series anomaly detection: a survey. ACM Comput. Surv. **57**(1), 1–42 (2024). https://doi.org/10.1145/3691338

Training for Mixed-Precision Integer Weights, Activations and Embeddings in BERT

Cédric Gernigon[1], Xavier Pillet[1,2], Anastasia Volkova[3(✉)], and Richard Dufour[1]

[1] Nantes Université, LS2N, UMR 6004, Nantes, France
{cedric.gernigon,xavier.pillet,richard.dufour}@univ-nantes.fr
[2] Valeuriad, Nantes, France
[3] Inria, INSA Lyon, CITI, UMR 3720, Villeurbanne, France
anastasia.volkova@inria.fr

Abstract. Pre-trained language models (LMs) deliver strong performance across a wide range of Natural Language Processing (NLP) tasks but remain costly to deploy on embedded devices due to their high memory and compute requirements. A widely used strategy for adapting LMs to resource-constrained devices is aggressive quantization. At low bit-widths, mixed-precision schemes, where different components of the model use different numerical precisions, offer an effective balance between compression and accuracy. In this work, we evaluate the impact of mixed-precision quantization for inference on the BERT language model. Unlike prior studies that often overlook activation quantization, our evaluation systematically explores mixed-precision configurations for both weights and activations. We also examine the effects of quantizing the embedding layer, which is commonly limited to token-weight quantization. Evaluated on the SQuAD and GLUE benchmarks, our approach achieves substantial reductions in memory and computational cost without sacrificing accuracy.

Keywords: Quantization · Mixed-precision · Inference acceleration · Embedding layer quantization · Quantization-Aware Training

1 Introduction

Pre-trained language models such as BERT [8] have achieved significant improvements across a wide range of NLP tasks [22,28], but these major gains come at the cost of massive parameter counts and high compute. For example, $\text{BERT}_{\text{BASE}}$ has 109M parameters (418MB memory size) and requires billions of operations. The need for real-time inference, along with privacy and security concerns, has made the efficient deployment of these models on edge devices a major challenge. Due to the limited resources, including energy, memory, and computational power of these devices, model compression is essential for their implementation. Various model compression methods have been successfully applied to BERT, such as pruning [12], knowledge distillation [24], and quantization [36].

M. Kowalczyk and C. Monière (Eds.): DASIP 2026, LNCS 16512, pp. 132–144, 2026.
https://doi.org/10.1007/978-3-032-23871-9_11

Quantization reduces model size and computational cost by representing weights and activations at lower bit-widths, potentially altering the underlying arithmetic. Transformer-based models can be uniformly quantized to 8-bit integers [36], or to ternary [37] or binary [1] representations with limited accuracy degradation. Advanced techniques [9,13,29,33] further increase compression by assigning different bit-widths across (sub)layers. Identifying optimal mixed-precision configurations that minimize accuracy loss remains challenging. Optimization-based methods have emerged as an effective approach, providing a favorable trade-off between compression efficiency and computational cost. Previous studies have predominantly focused on weight quantization, as linear layers constitute the majority of model memory. Nevertheless, deployed models execute all parameters and computations in low precision, highlighting a gap between software-modeled performance and actual hardware accuracy.

In this paper, we push mixed-precision quantization beyond weights-only to include activations and all embedding parameters in BERT using AdaQAT [11]. AdaQAT is a gradient-based optimization framework for mixed-precision quantization that learns bit-widths to jointly quantize weights and activations during training, utilizing a Quantization-Aware Training (QAT) strategy. Its key lies in the use of relaxed fractional bit-widths, which are continuously optimized via gradient descent but discretized during both forward and backward passes. We propose a per-layer mixed-precision approach permitting a fine-grained optimization. We evaluate our approach on the SQuAD dataset, enabling direct comparison with state-of-the-art methods, and on GLUE benchmark to investigate the impact of mixed-precision quantization over different language tasks. Our experiments demonstrate that the proposed method achieves substantial reductions in memory and computational overhead for $\text{BERT}_{\text{BASE}}$ while maintaining task accuracy. In addition, we examine the effect of quantizing the embedding layer's weights on model accuracy. While conventional approaches often focus exclusively on token embeddings, our findings demonstrate that *all* embedding weights can be quantized with minimal impact on performance. For instance, on the SQuAD dataset, quantizing the full embedding layer to 4 bits or below preserves accuracy while halving the memory requirements of the embedding layer. The tool and experiments are open-source and available[1].

This article is structured as follows. Section 2 reviews related work on QAT and mixed-precision methods. Section 3 describes our methodology, including the integration of knowledge distillation into the mixed-precision flow. Finally, Sect. 4 presents experimental results, analyzing the performance of mixed-precision quantization applied to the $\text{BERT}_{\text{BASE}}$ model and the impact of embedding layer quantization.

2 Related Work

There are two primary approaches to quantization of DNNs. The first, Post-Training Quantization (PTQ), applies quantization after training and is fast to

[1] The code is at https://github.com/cgernigon/AdaQAT.

deploy, but can incur non-negligible accuracy loss at very low bit-widths [2]. In contrast, Quantization-Aware Training (QAT), while slower and requiring access to training data, generally leads to better results.

Quantization-Aware Training (QAT) has been widely studied to train CNNs with small bit-widths. QAT methods stem from pioneering work on binary neural networks [6,7]. At its core, a QAT method consists of using a quantized version of the network during training in both forward and backward passes, while performing updates on full-precision copies of the network parameters. These full-precision parameters are then quantized to be used in the next iteration. A crucial aspect is how to perform backpropagation through quantized variables (parameters and activations). In the binary case, this was done using a so-called Straight-Through Estimator (STE) [3], and this approach was later extended to cover larger bit-widths [39], while also applying quantization to gradient signals. To further improve the accuracy of quantized networks, the STE idea can also be used to learn the parameters of uniform quantizers, such as scaling factors and bias terms for weight quantization [4,10], and in the case of ReLU-based activations, clipping parameters [5].

QAT has been successfully applied to LLMs, with BERT being trained in 8-bit integer QAT with minimal accuracy loss using STE-based approaches [36]. Quantization with lower bit-widths has also been explored, including ternary [37] and binary [1,21] quantized BERT. [26] proposed to replace the STE-based gradient for updating the quantization scale with an MSE-based gradient strategy. The SQuAT [30] method alternates training between the sharpness objective and the step-size objective, allowing the model to learn the most appropriate parameter update magnitude to reach convergence close to flat minima.

Mixed-Precision for Inference aims to reduce bit-widths at a finer granularity, typically at the (sub)layer level, which is challenging without automated configuration methods. Several approaches have been proposed to address this problem.

Search-based methods like HAQ [29] rely on reinforcement learning with hardware (latency & energy) feedback in the agent, whereas neural architecture search works like DNAS [31] uses gradient-based information. The major downside of using them is that they require significant time and computational resources. AQ-BERT [38] adopts a Network Architecture Search approach to set different precision for each neuron sub-group of BERT.

Much faster results can be obtained using metric-based methods. For example, HAWQ [9] uses Hessian spectrum information at each layer to assign precisions; this approach was used in Q-BERT [25]. Methods like in [18,34] rely on linear programming models, while [16] encourages quantization that leads to reduced sharpness in the task loss function. These approaches are fast but can be sub-optimal compared to more global optimization strategies [13].

Optimization-based approaches formulate the bit-width assignment as an optimization problem, with the main challenge being how to handle the fact that the loss is non-differentiable w.r.t. the bit-widths. Methods like FracBits [33] and BitPruning [20] use fractional bit-widths and linear interpolation during the forward path, whereas SDQ [13] is based on stochastic quantization, but seems limited to weight quantization. These methods work well in fine-tuning scenarios, but are unstable or do not work when training from scratch.

AdaQAT belongs to the optimization-based family and offers a flexible QAT framework that learns relaxed (fractional) bit-widths via gradient-based optimization. This paper extends AdaQAT to LLMs and improves its applicability to BERT, including joint treatment of activations and embeddings.

3 Mixed-Precision Exploration with AdaQAT

For simplicity, we start by presenting AdaQAT in the context of learning only one bit-width for all weights in the network and similarly for activations. We then extend the method to per-layer bit-width allocation.

3.1 Objective Function

To learn the bit-widths of the uniform quantizers for both weights and activations, we use two real-valued variables $N_{\mathbf{w}}$ and $N_{\mathbf{a}}$, respectively. The actual integer bit-widths of the quantized network will be $\lceil N_{\mathbf{w}} \rceil$ and $\lceil N_{\mathbf{a}} \rceil$.

We use a regularization approach to model the loss function that takes into account the cost of a particular bit-width configuration, which corresponds to:

$$\mathcal{L}_{\text{Total}} = \mathcal{L}_{\text{Task}}\left(\mathbf{Y}_{\text{U,U}}^{(L)}; \mathbf{T}\right) + \lambda \mathcal{L}_{\text{Hard}}\left(\lceil N_{\mathbf{w}} \rceil, \lceil N_{\mathbf{a}} \rceil\right), \quad (1)$$

where $\lambda > 0$ is a regularization hyper-parameter between the main task loss $\mathcal{L}_{\text{Task}}$ (*i.e.*, the usual loss for the task that needs to be addressed) and the regularizing hardware loss $\mathcal{L}_{\text{Hard}}$ that measures the complexity of the model (in this case in terms of hardware cost). Here, $\mathbf{Y}_{\text{U,U}}^{(L)}$ represents the output of the L-layer quantized network we want to optimize, where its weights have been quantized to $\lceil N_{\mathbf{w}} \rceil$ bits (the first U in the superscript notation) and its activations to $\lceil N_{\mathbf{a}} \rceil$ bits (the second U in the superscript) on a current minibatch of data and $\mathbf{T}$ is the expected output.

FracBits [33] has reviewed various methods used to model the hardware cost of arithmetic precision choices for weights and activations. They argue in favor of memory size if only targeting weight quantization, and BitOPs for joint weight and activation quantization. As an example, for a linear layer, the BitOPs metric corresponds to

$$\text{BitOPs}(l) = \lceil N_{\mathbf{w}} \rceil \lceil N_{\mathbf{a}} \rceil |w_l|, \quad (2)$$

where l is a linear layer and $|w_l|$ denotes the number of weights in the linear layer l. Thus, the overall BitOPs of a model corresponds to the sum of each linear layer's BitOPs.

In our particular case, since we are using one bit-width per weights and one per activations, the overall BitOPs hardware cost will be linear in $\lceil N_{\mathbf{w}} \rceil \lceil N_{\mathbf{a}} \rceil$, and we can simplify the regularization term to

$$\mathcal{L}_{\text{Hard}}\left(\lceil N_{\mathbf{w}} \rceil, \lceil N_{\mathbf{a}} \rceil\right) = \lceil N_{\mathbf{w}} 13 - -14 \rceil \lceil N_{\mathbf{a}} \rceil . \tag{3}$$

Using Eq. 3 will impact the value we need to consider for λ, which will be smaller than when integrating the scaling factors specific to each layer (determined by the layer type and shape) directly into $\mathcal{L}_{\text{Hard}}$.

3.2 Bit-Width Gradients and Parameter Updates

Since the task loss $\mathcal{L}_{\text{Task}}$ is not directly differentiable with respect to the bit-width parameters, we use finite difference approximations as

$$\begin{aligned} \frac{\partial \mathcal{L}_{\text{Task}}}{\partial N_{\mathbf{w}}} &\approx \mathcal{L}_{\text{Task}}(\mathbf{Y}^{(L)}_{\text{U,U}}; \mathbf{T}) - \mathcal{L}_{\text{Task}}(\mathbf{Y}^{(L)}_{\text{D,U}}; \mathbf{T}), \\ \frac{\partial \mathcal{L}_{\text{Task}}}{\partial N_{\mathbf{a}}} &\approx \mathcal{L}_{\text{Task}}(\mathbf{Y}^{(L)}_{\text{U,U}}; \mathbf{T}) - \mathcal{L}_{\text{Task}}(\mathbf{Y}^{(L)}_{\text{U,D}}; \mathbf{T}). \end{aligned} \tag{4}$$

The U (up) and D (down) subscripts in the $\mathbf{Y}^{(L)}$ outputs denote how the weights (first subscript) and activations (second subscript) have been rounded.

The gradients of the total loss 1 w.r.t. the bit-widths are

$$\frac{\partial \mathcal{L}_{\text{Total}}}{\partial N_{\mathbf{x}}} \approx \frac{\partial \mathcal{L}_{\text{Task}}}{\partial N_{\mathbf{x}}} + \lambda \frac{\partial \mathcal{L}_{\text{Hard}}}{\partial \lceil N_{\mathbf{x}} \rceil}, \tag{5}$$

which are then used to update the fractional bit-width parameters. The gradient descent rule that does this takes the form

$$N_{\mathbf{x}}^{+} = N_{\mathbf{x}} - \eta_{\mathbf{x}} \frac{\partial \mathcal{L}_{\text{Total}}}{\partial N_{\mathbf{x}}}, \tag{6}$$

with $\mathbf{x} \in \{\mathbf{w}, \mathbf{a}\}$, $N_{\mathbf{x}}^{+}$ the new bit-width values at the next iteration, and $\eta_{\mathbf{x}} > 0$ the corresponding learning rates.

To summarize, during the *forward propagation phase*, the weights and activations are quantized according to the learned integer bit-widths, $\lceil N_{\mathbf{w}} \rceil$ and $\lceil N_{\mathbf{a}} \rceil$, respectively, and to the quantization policy. In the *gradient estimation phase*, the bit-width gradients are approximated according to Eqs. 4 and 5, while in the *parameter update phase*, we use Eq. 6 to update the bit-width parameters. The rest of the network and quantizer parameters are updated in parallel using standard optimizers, with their separate hyperparameters.

When $N_{\mathbf{w}}$ and $N_{\mathbf{a}}$ reach their optimized values, further decreases cause a steep increase in the task loss $\mathcal{L}$Task and consequently $\mathcal{L}$Total. Their gradient estimates in Eq. 5 then become negative, and the update rule in Eq. 6 starts increasing $N_{\mathbf{w}}$ and $N_{\mathbf{a}}$, producing an oscillatory pattern observed in several STE-based gradient approximation methods [15,17,19]. We monitor the oscillations, and once they exceed a threshold (empirically set to 10), we fix the bit-widths to $\lceil N_{\mathbf{w}} \rceil$ and $\lceil N_{\mathbf{a}} \rceil$, and continue the quantization process in the standard fashion.

3.3 Per-Layer Extension

Extending the method to layer-wise precision configurations requires finding appropriate bit-width variables for each layer of the model such that, for an L-layer model, $\left\lceil N_{\mathbf{w}}^{(i)} \right\rceil$ and $\left\lceil N_{\mathbf{a}}^{(i)} \right\rceil$ are the bit-widths for layer $i \in [\![1, L]\!]$. This problem is non-trivial, since the computational cost of the gradient estimation phase would grow at least linearly with the number of bit-widths to learn. Instead, it is more cost-effective to focus on updating the bit-width of only one layer's weights or activations at a time, potentially switching the layer to update at each iteration. The overall approach can thus be viewed as a quantized version of a *block coordinate descent* [27] update rule. Block coordinate optimization methods update only a group of parameters at a time and have emerged as viable approaches to solve complex nonconvex optimization problems, like those involving "big data" (see [32] for instance and the references therein).

Careful attention must be paid when selecting which bit-widths to update at a given iteration. The most straightforward approach is to take a bit-width parameter at random, but this turns out to be highly suboptimal in our case. Greedy-type approaches offer better results: at each iteration, update (*e.g.*, decrease) the bit-width that is most likely to degrade accuracy the least on the current training batch.

Weights and Activations. Similarly to the per-network bit-width learning context, one must decide if weight and activation bit-widths should be updated simultaneously or in sequence. Whereas in the per-network setting model accuracy is not sensitive to this choice (simultaneous and in-sequence updates tend to behave analogously), at finer granularities (*e.g.*, layer or channel) this is not the case. Taking bit-widths separately allows more freedom in exploring the search space. It is also generally the case that activations are more sensitive to perturbations than weights, meaning that model accuracy is less impacted by focusing on weights bit-width learning first and then moving on to activation bit-widths. The fact that activation values depend on the weights (and inputs) also makes it more natural to start with weights.

Bit-width Update Selection Metrics. Choosing to always update the bit-width that leads to the least significant reduction (*i.e.*, with the smallest cost) is a sensible approach. Accordingly, we adopted a Mean Squared Error (MSE) strategy to select which bit-width to update in the subsequent block coordinate descent iteration. For two quantized tensors of the same shape $\widetilde{\mathbf{X}}_1$ and $\widetilde{\mathbf{X}}_2$ (*i.e.*, either weights or activations, before and after the corresponding bit-width parameter has been updated), the cost is their mean squared error:

$$\mathrm{MSE}\left(\widetilde{\mathbf{X}}_1, \widetilde{\mathbf{X}}_2\right) := \left\|\widetilde{\mathbf{X}}_1 - \widetilde{\mathbf{X}}_2\right\|_F^2 / \mathrm{size}(\widetilde{\mathbf{X}}_1), \tag{7}$$

where $\mathrm{size}(\widetilde{\mathbf{X}}_1)$ is the number of elements in $\widetilde{\mathbf{X}}_1$ and $\|\cdot\|_F$ is the Frobenius norm.

4 Experiments and Discussion

This section provides an experimental evaluation of our PyTorch-based framework, compared to the state-of-the-art. We evaluate model efficiency using memory size and the BitOPs metric (2). As state-of-the-art mixed-precision models do not report per-layer bit-widths, we approximate BitOPs via memory size. Given that activations are uniform across layers, they can be factorized, reducing the calculation to the memory size of weights (excluding embedding weights).

Table 1. Comparison with state-of-the-art quantization methods on SQuAD.

Model	Bit-width	Size	BitOPs	SQuAD v1.1		SQuAD v2.0	
	W-E-A	(MB)	($\times 10^9$)	EM	F_1-score	EM	F_1-score
$BERT_{BASE}$	fp32	417.6	/	81.5	88.7	74.5	77.7
TernaryBERT	2-2-8	28.0	1.368	80.1	87.5	73.3	76.6
BinaryBERT	1-1-4	16.5	0.342	79.3	87.2	72.5	75.4
Q-BERT	MP-8-8	53.2	≈ 1.80	79.85	87.49	/	/
AQ-BERT	MP-8-8	55.0	≈ 1.92	79.85	87.00	/	/
AdaQAT	2.08-4-3.93	29.2	0.677	/	/	75.2	77.9
	2.06-4-3.96	35.9	0.701	81.07	88.71	/	/

$$\text{BitOPs}(M) = N_{\mathbf{a}} \sum_{i=0}^{L-1} \left\lceil N_{\mathbf{w}}^{(i)} \right\rceil |w_l| = N_{\mathbf{a}} \times \text{memsize}(W) \tag{8}$$

where M is an L-layer model and memsize(W) is computed in MB.

All tables report bit-widths in the WâĂŞEâĂŞA format, corresponding to linear-layer weights, word embeddings, and activations. When zooming into embedding layers, T-S-P refers to token, segment and position. For state-of-the-art methods without precise configurations, MP denotes variable-length layers. Size includes embedding weights unless stated otherwise. Depending on the task, we report exact match (EM) and F_1 scores.

Quantization Procedures. We quantize the weights and the inputs (activations) from all Transformer layers. For a fair comparison with state-of-the-art, only word embeddings are quantized in the sections on mixed-precision. The embedding bit-width is also not included in the mixed-precision search space. The embedding layer quantization is explored in detail in Sect. 4.3. Biases in linear layers, softmax, layer normalization, and the final task-specific layer remain unquantized. LSQ [10] is used to learn quantization ranges for both weights and activations. We use the fp32 BERT fine-tuned on the downstream task to initialize our quantized model.

Benchmarks. We evaluate our models on SQuAD and GLUE. SQuAD v1.1 [22] contains over 100,000 question-answer pairs based on Wikipedia, and v2.0 [23] adds over 50,000 unanswerable questions. GLUE [28] comprises nine natural language understanding tasks spanning diverse datasets, difficulties, and text genres. These benchmarks assess performance across a broad range of reading comprehension and linguistic phenomena.

4.1 Comparison of AdaQAT with SotA

Table 1 shows the results of applying AdaQAT on SQuAD datasets compared to other methods from the literature. The first line shows the floating-point baseline result, whereas the second group of lines showcases *static* methods, where activations are not quantized and weights are quantized uniformly. The BitOPs for Q-BERT and AQ-BERT were not provided by the authors, hence we approximate them using the given memory size and the model linear weights.

The third group of lines corresponds to mixed-precision methods where the weight bit-width is learned and the activations are quantized uniformly to a manually set bit-width. The last groups of lines present our mixed-precision $\text{BERT}_{\text{BASE}}$ results on SQuAD. EM and F_1-score are similar to or slightly better than the fp32 baseline. Compared to the other mixed-precision models Q-BERT and AQ-BERT, considering activations significantly reduces the computational cost of the model. Our configurations also significantly reduced the size required to store the weights in memory, while achieving better accuracy. Although mixed-precision models require more memory than binary or ternary models, they offer greater accuracy closer to the baseline. The computational cost of activations is also not negligible, with much higher BitOPs for TernaryBERT despite a smaller memory size for weights compared to our mixed-precision models.

4.2 Mixed-Precision for Diverse Language Tasks with GLUE Dataset

Table 2 presents the state-of-the-art results for quantized $\text{BERT}_{\text{BASE}}$ models on the GLUE benchmark. Note that the mixed-precision approaches, up to our knowledge, have considered only a subset of GLUE datasets/tasks. Table 3 presents our mixed-precision results on the full GLUE benchmark.

Compared to Q-BERT, our approach demonstrates clearly superior results in both accuracy and size. By including the embedding and activation layer quantization into the optimization objective, AdaQAT usually maintains the baseline accuracy drop within 1% while dividing the memory footprint by roughly 11 compared to the fp32 baseline, and by 1.5 compared to Q-BERT. This demonstrates the robustness of BERT towards full model quantization.

Extremely low but uniform precision models, TernaryBERT and BinaryBERT, perform very well on most of the tasks but suffer on datasets that are sensitive to quantization, e.g. CoLa. With AdaQAT one can automatically find mixed-precision configurations that have slightly larger configurations that add the necessary bit-widths at appropriate layers. For instance, for CoLA, we

Table 2. State-of-the-Art quantization results on the GLUE benchmark.

Model	Bit-width W-E-A	Size (MB)	BitOPs ($\times 10^9$)	MNLI -m/mm	QQP	QNLI	SST-2	CoLA	STS-B	MRPC	RTE
BERT$_{\text{BASE}}$	fp32	417.6	/	84.57/84.46	91.39	91.34	92.32	59.67	90.14	86.28	69.93
Q-BERT	2-8-8	43.1	1.368	76.56/77.02	/	/	84.63	/	/	/	/
	MP-8-8	53.2/53.2	$\approx$ 1.80	83.51/83.55	/	/	92.55	/	/	/	/
	MP-8-8	46.1/48.1	$\approx$ 1.35/1.48	81.75/82.29	/	/	92.08	/	/	/	/
TernaryBERT	2-2-8	28.0	1.368	83.0/82.2	88.4	90.0	92.9	47.8	84.3	87.5	68.4
BinaryBERT	1-1-4	16.5	0.342	83.9/84.2	91.2	90.9	92.3	44.4	87.2	83.3	65.3

obtained that a configuration 2.41-4-4.30 limits the accuracy drop to only 1%, while the TernaryBERT 2-2-8 configuration suffers a roughly 12% accuracy loss compared to the baseline. This non-trivial solution of decreasing activation bit-widths while non-homogeneously increasing the formats of weights and embeddings is impossible to find by hand but requires an optimization-based approach.

Table 3. Our mixed-precision results with AdaQAT on the GLUE benchmark.

	MNLI -m/mm	QQP	QNLI	SST-2	CoLA	STS-B	MRPC	RTE
Accuracy (%)	84.66/84.38	90.96	91.72	92.43	58.02	89.73	86.50	69.10
Bit-width W-E-A	2.25-4-4.11	2.16-4-4.03	2.18-4-4.14	2.16-4-4.07	2.41-4-4.30	2.27-4-4.21	2.24-4-4.18	2.47-4-4.34
Size (MB)	36.67	36.23	36.75	37.11	39.77	37.70	37.04	40.28
BitOPs ($\times 10^9$)	0.756	0.737	0.773	0.781	0.902	0.815	0.792	0.925

4.3 Quantization Effects on the Embedding Layer

The weights of the embedding layer account for 21% of the total parameters in the BERT$_{\text{BASE}}$ model. Of this 21%, 98% are token weights. For this reason, in SotA approaches, token weights are typically included in quantization exploration, while segment and position weights are not quantized as they are considered negligible. While this is a reasonable argument, this perspective is inappropriate when considering compression for model deployment on resource-constrained devices.

Q-BERT [25] examines the impact of quantizing word embeddings and position embeddings in the BERT$_{\text{BASE}}$ model. The authors argue that the embedding layer is more sensitive to quantization than the linear layer, with performance drops of up to 10% on tasks like SST-2, MNLI, and CoNLL-03, and over 20% on SQuAD v1.1 when using 4-bit embedding quantization. Additionally, they find that position embeddings are more sensitive to quantization, leading to an extra 2% performance degradation compared to quantizing word embeddings alone.

GOBO [35] observed different behaviors, suggesting that it is nevertheless possible to lower the bit-width of the embeddings without significantly impacting accuracy, which we confirm with our experiments as well.

Table 4 shows the size of the embedding layer, model accuracy across multiple datasets and different quantization configurations. These results indicate that it is actually possible to quantize all weights in the embedding layer to 2-bit while maintaining the model accuracy drop below 2% on the evaluated datasets. The memory footprint of the embedding layer is reduced by a factor of 16 compared to the single-precision model and by a factor of 4.2 compared to the approach generally used in state-of-the-art work, i.e. quantizing only the token weights.

Table 4. Evaluation of the quantization effects of embedding weights on various tasks. CR is the embedding weights compression ratio w.r.t. fp32 baseline.

Model	Bit-width	Embedding	Embedding	CR	MNLI	SST-2	SQuAD v1.1		SQuAD v2.0	
	W-A	T-S-P	size (MB)	×	-m/mm		EM	F_1-score	EM	F_1-score
$\text{BERT}_{\text{BASE}}$	fp32	fp32	95.32	1.00	84.57/84.46	92.32	81.41	88.61	74.45	77.64
Q-BERT	8–8	8-fp32-8	23.84	4.00	83.83/83.91	92.88	/	88.47	/	/
		4-fp32-8	12.12	7.86	82.91/83.67	91.74	/	87.55	/	/
		8-fp32-4	23.64	4.03	82.84/82.25	89.11	/	72.38	/	/
		4-fp32-4	11.92	8.00	78.08/78.96	85.55	/	61.70	/	/
AdaQAT	8–8	8-fp32-fp32	25.01	3.81	84.57/84.46	92.89	83.59	90.56	75.17	78.17
		4-fp32-fp32	13.30	7.17	84.95/84.72	92.66	83.54	90.55	75.20	78.24
		8-fp32-8	23.84	4.00	85.05/84.78	92.55	83.55	90.57	75.27	78.20
		4-4-4	11.91	8.00	85.02/84.43	92.78	83.54	90.56	75.28	78.23
		2-2-2	5.96	16.0	84.97/84.64	92.66	83.60	90.47	75.38	78.29
	2–4	4-fp32-fp32	13.30	7.17	83.89/84.14	91.17	81.58	89.00	74.98	76.86
		4-fp32-4	11.92	8.00	83.76/83.60	91.17	81.47	89.00	74.80	77.60
		4-4-4	11.91	8.00	83.78/84.11	91.17	81.51	89.00	74.43	77.44
		2-2-2	5.96	16.0	83.79/84.30	90.60	81.36	88.91	74.11	77.14

5 Conclusion and Future Work

In this paper, we improve the efficiency of BERT-based models via AdaQAT, a mixed-precision algorithm that considers weights, activations, and embedding layers. Our mixed-precision $\text{BERT}_{\text{BASE}}$ uses low bit-width (< 8) fixed-point representations across linear and activation layers, often preserving baseline accuracy while reducing computational cost by over 2× and memory size by 1.5× compared to weight-only quantization approaches. We further evaluate quantizing all embedding weights, showing that 2-bit representations maintain accuracy within 2% of the baseline while reducing the embedding memory footprint by 16× versus fp32 and 4.2× versus prior work.

Future work includes applying mixed-precision to smaller models such as TinyBERT [14] and DistilBERT [24], which are more sensitive to quantization, as well as exploring domain-specific models and datasets where non-trivial mixed-precision assignments may yield efficient configurations.

Acknowledgment. This work was performed using HPC resources from GENCI-IDRIS (Grant 2025-AD011016060). This work has received French government support granted to the CominLabs Excellence Laboratory and managed by the National Research Agency in the "Investing for the Future" program under reference ANR-10-LABX-07-01. This work has been partially funded by the French National Research Agency project PEPR AI under the reference ANR-23-PEIA-0010.

References

1. Bai, H., et al.: Binarybert: pushing the limit of Bert quantization. arXiv preprint arXiv:2012.15701 (2020)
2. Banner, R., Nahshan, Y., Soudry, D.: Post training 4-bit quantization of convolutional networks for rapid-deployment. Adv. Neural Inf. Process. Syst. **32** (2019)
3. Bengio, Y., Léonard, N., Courville, A.: Estimating or propagating gradients through stochastic neurons for conditional computation. arXiv preprint arXiv:1308.3432 (2013)
4. Bhalgat, Y., Lee, J., Nagel, M., Blankevoort, T., Kwak, N.: LSQ+: improving low-bit quantization through learnable offsets and better initialization. In: Proceedings of the IEEE/CVF Conference on Computer Vision and Pattern Recognition Workshops, pp. 696–697 (2020)
5. Choi, J., Wang, Z., Venkataramani, S., Chuang, P.I.J., Srinivasan, V., Gopalakrishnan, K.: PACT: parameterized clipping activation for quantized neural networks. arXiv preprint arXiv:1805.06085 (2018)
6. Courbariaux, M., Bengio, Y., David, J.P.: BinaryConnect: training deep neural networks with binary weights during propagations. In: Advances in Neural Information Processing Systems 28 – NIPS 2015. NIPS'15, vol. 2, pp. 3123–3131. MIT Press (2015)
7. Courbariaux, M., Hubara, I., Soudry, D., El-Yaniv, R., Bengio, Y.: Binarized neural networks: training deep neural networks with weights and activations constrained to +1 or −1. arXiv preprint arXiv:1602.02830 (2016)
8. Devlin, J., Chang, M.W., Lee, K., Toutanova, K.: Bert: pre-training of deep bidirectional transformers for language understanding. In: Proceedings of NAACL-HLT. vol. 1, p. 2 (2019)
9. Dong, Z., Yao, Z., Gholami, A., Mahoney, M.W., Keutzer, K.: HAWQ: hessian aware quantization of neural networks with mixed-precision. In: Proceedings of the IEEE/CVF International Conference on Computer Vision, pp. 293–302 (2019)
10. Esser, S.K., McKinstry, J.L., Bablani, D., Appuswamy, R., Modha, D.S.: Learned step size quantization. arXiv preprint arXiv:1902.08153 (2019)
11. Gernigon, C., Filip, S.I., Sentieys, O., Coggiola, C., Bruno, M.: Adaqat: adaptive bit-width quantization-aware training. In: 2024 IEEE 6th International Conference on AI Circuits and Systems (AICAS), pp. 442–446. IEEE (2024)
12. Gordon, M.A., Duh, K., Andrews, N.: Compressing Bert: studying the effects of weight pruning on transfer learning. arXiv preprint arXiv:2002.08307 (2020)

13. Huang, X., et al.: SDQ: stochastic differentiable quantization with mixed precision. In: International Conference on Machine Learning, pp. 9295–9309. PMLR (2022)
14. Jiao, X., et al.: Tinybert: distilling Bert for natural language understanding. arXiv preprint arXiv:1909.10351 (2019)
15. Kuzmin, A., Van Baalen, M., Ren, Y., Nagel, M., Peters, J., Blankevoort, T.: Fp8 quantization: the power of the exponent. Adv. Neural. Inf. Process. Syst. **35**, 14651–14662 (2022)
16. Liu, J., Cai, J., Zhuang, B.: Sharpness-aware quantization for deep neural networks. arXiv preprint arXiv:2111.12273 (2021)
17. Liu, S.Y., Liu, Z., Cheng, K.T.: Oscillation-free quantization for low-bit vision transformers. In: International Conference on Machine Learning, pp. 21813–21824. PMLR (2023)
18. Ma, Y., et al.: OMPQ: orthogonal mixed precision quantization. arXiv preprint arXiv:2109.07865 (2021)
19. Nagel, M., Fournarakis, M., Bondarenko, Y., Blankevoort, T.: Overcoming oscillations in quantization-aware training. In: International Conference on Machine Learning, pp. 16318–16330. PMLR (2022)
20. Nikolić, M., et al.: BitPruning: learning bitlengths for aggressive and accurate quantization. arXiv preprint arXiv:2002.03090 (2020)
21. Qin, H., et al.: Bibert: accurate fully binarized Bert. arXiv preprint arXiv:2203.06390 (2022)
22. Rajpurkar, P.: Squad: 100,000+ questions for machine comprehension of text. arXiv preprint arXiv:1606.05250 (2016)
23. Rajpurkar, P., Jia, R., Liang, P.: Know what you don't know: unanswerable questions for squad. arXiv preprint arXiv:1806.03822 (2018)
24. Sanh, V., Debut, L., Chaumond, J., Wolf, T.: Distilbert, a distilled version of Bert: smaller, faster, cheaper and lighter. arXiv preprint arXiv:1910.01108 (2019)
25. Shen, S., et al.: Q-Bert: hessian based ultra low precision quantization of Bert. In: Proceedings of the AAAI Conference on Artificial Intelligence. vol. 34, pp. 8815–8821 (2020)
26. Tang, H., Zhang, X., Liu, K., Zhu, J., Kang, Z.: MKQ-Bert: quantized Bert with 4-bits weights and activations. arXiv preprint arXiv:2203.13483 (2022)
27. Tseng, P.: Convergence of a block coordinate descent method for nondifferentiable minimization. J. Optim. Theory Appl. **109**, 475–494 (2001)
28. Wang, A.: Glue: a multi-task benchmark and analysis platform for natural language understanding. arXiv preprint arXiv:1804.07461 (2018)
29. Wang, K., Liu, Z., Lin, Y., Lin, J., Han, S.: HAQ: hardware-aware automated quantization with mixed precision. In: IEEE/CVF Conference on Computer Vision and Pattern Recognition (CVPR) (2019). https://doi.org/10.1109/CVPR.2019.00881
30. Wang, Z., Li, J.B., Qu, S., Metze, F., Strubell, E.: Squat: sharpness-and quantization-aware training for Bert. arXiv preprint arXiv:2210.07171 (2022)
31. Wu, B., Wang, Y., Zhang, P., Tian, Y., Vajda, P., Keutzer, K.: Mixed precision quantization of convnets via differentiable neural architecture search. arXiv preprint arXiv:1812.00090 (2018)
32. Xu, Y., Yin, W.: A globally convergent algorithm for nonconvex optimization based on block coordinate update. J. Sci. Comput. **72**(2), 700–734 (2017)
33. Yang, L., Jin, Q.: FracBits: mixed precision quantization via fractional bit-widths. In: Proceedings of the AAAI Conference on Artificial Intelligence, pp. 10612–10620 (2021)
34. Yao, Z., et al.: HAWQ-V3: dyadic neural network quantization. In: International Conference on Machine Learning, pp. 11875–11886. PMLR (2021)

35. Zadeh, A.H., Edo, I., Awad, O.M., Moshovos, A.: Gobo: quantizing attention-based NLP models for low latency and energy efficient inference. In: 2020 53rd Annual IEEE/ACM International Symposium on Microarchitecture (MICRO), pp. 811–824. IEEE (2020)
36. Zafrir, O., Boudoukh, G., Izsak, P., Wasserblat, M.: Q8bert: quantized 8bit Bert. In: 2019 Fifth Workshop on Energy Efficient Machine Learning and Cognitive Computing-NeurIPS Edition (EMC2-NIPS), pp. 36–39. IEEE (2019)
37. Zhang, W., et al.: Ternarybert: distillation-aware ultra-low bit bert. arXiv preprint arXiv:2009.12812 (2020)
38. Zhao, C., Hua, T., Shen, Y., Lou, Q., Jin, H.: Automatic mixed-precision quantization search of Bert. arXiv preprint arXiv:2112.14938 (2021)
39. Zhou, S., Wu, Y., Ni, Z., Zhou, X., Wen, H., Zou, Y.: DoReFa-Net: training low bitwidth convolutional neural networks with low bitwidth gradients. arXiv:1606.06160 (2016)

Secure and Dependable Embedded Systems

ARMOR: Accelerator Runtime Monitoring and cOntrolled Identity enfoRcement

Aya Jendoubi[1](✉), Jean-Christophe Prevotet[1], Philippe Tanguy[2], and Pascal Cotret[3]

[1] INSA Rennes - CNRS, IETR, UMR 6164, Rennes, France
{aya.jendoubi,jean-christophe.prevotet}@insa-rennes.fr
[2] Université Bretagne-Sud - Lab-STICC, UMR CNRS 6285, Lorient, France
philippe.tanguy@univ-ubs.fr
[3] ENSTA - Lab-STICC, UMR CNRS 6285, Brest, France
pascal.cotret@ensta.fr

Abstract. Dynamically reconfigurable hardware offers significant adaptability and performance advantages for modern System-on-Chip (SoC) architectures, but also enlarges the system's attack surface. Previous work has shown that reconfigurable accelerators can exploit weaknesses in Input Output Memory Management Unit (IOMMU) to perform unauthorized Direct Memory Accesss (DMAs), revealing limitations in current I/O protection mechanisms. This paper presents ARMOR, a hardware mechanism designed to ensure the trustworthy integration of reconfigurable accelerators within IOMMU-enabled SoCs. ARMOR enforces secure device identity verification and provides runtime supervision of accelerator activity to detect and prevent abnormal or malicious behavior. Implemented in a RISC-V environment and validated through simulation, it effectively prevents I/O-based attacks while maintaining system compatibility and low performance overhead.

Keywords: IOMMU · Hardware Accelerator · Dynamic Reconfiguration · Wrapper · System on Chip

1 Introduction

Recent progress in hardware architectures has been driven by modern software applications' demands for enhanced functionalities, processing power, and communication capabilities. This has led to the integration of dedicated hardware units, such as Hardware Accelerators (HAs) and coprocessors, which operate independently of the CPU, improving system efficiency by offloading specific tasks and enabling greater parallelism. The introduction of Dynamic Partial Reconfigurations (DPRs) further amplifies this adaptability by allowing hardware modules to be reconfigured at runtime, catering to evolving application needs without halting the entire system. As systems grow more complex, seamless collaboration between hardware and software layers is essential for optimiz-

M. Kowalczyk and C. Monière (Eds.): DASIP 2026, LNCS 16512, pp. 147–158, 2026.
https://doi.org/10.1007/978-3-032-23871-9_12

ing performance while maintaining effective resource sharing across heterogeneous platforms. However, increased interconnectivity introduces new security challenges, particularly regarding I/O operations.

Modern heterogeneous System-on-Chips increasingly integrate DMA-enabled HAs to enhance computational throughput and allow for task specialization. DMA poses significant risks, as it allows peripheral devices to access system memory directly, potentially bypassing traditional CPU-based security. In these platforms, the IOMMU is responsible for enforcing spatial isolation by performing permission checks and translating device-generated virtual addresses into system physical addresses. However, in dynamically reconfigurable architectures, hardware functions can be instantiated, modified, or replaced at runtime, inherently broadening the attack surface and reducing the predictability of the hardware trust boundary as shown in [3,4,9]. Authors in [5] show that even more recent architectures of IOMMU (such as RISC-V IOMMU, which are extremely efficient and secure in static environments, can be bypassed as they assume the correctness of HA-supplied Device IDs and do not incorporate cryptographic or structural authentication to validate their provenance. This implicit trust creates a vulnerability in which adversaries may forge Identifiers (IDs) to escalate access privileges and bypass memory protection policies. Additionally, the shared communication substrate required to interconnect multiple HAs introduces contention-based threats, where malicious traffic amplification or interrupt storms can compromise system availability and degrade quality of service for benign components.

The remainder of this paper is organized as follows. Section 2 introduces the problem statement and the case study. Section 3 introduces the ARMOR hardware mechanism in detail, describing its architecture and security features. Section 4 reviews related works addressing similar challenges and highlights how ARMOR differentiates itself from existing solutions. Finally, Sect. 5 concludes the paper.

2 Case Study

In this work, we consider a heterogeneous RISC-V system in which multiple HAs are deployed within a Dynamically Reconfigurable Area (DRA) of an FPGA. All HAs communicate with system memory through a shared AXI interconnect and a single RISC-V IOMMU, as shown in Fig. 1.

In this architecture, communication between HAs, the AXI interconnect, and the IOMMU adheres to the AXI protocol, which employs a decoupled handshake mechanism based on the VALID and READY signaling scheme across its address, data, and response channels. This protocol enables high-throughput pipelined communication, wherein a transaction is acknowledged only upon the concurrent assertion of both signals. Each channel conveys control information, including transaction IDs and device IDs, which the IOMMU utilizes to enforce access isolation among HAs.

Each HA is allocated a distinct memory domain; however, as demonstrated in [5], a Malicious Hardware Accelerator (MHA) can exploit the lack of device

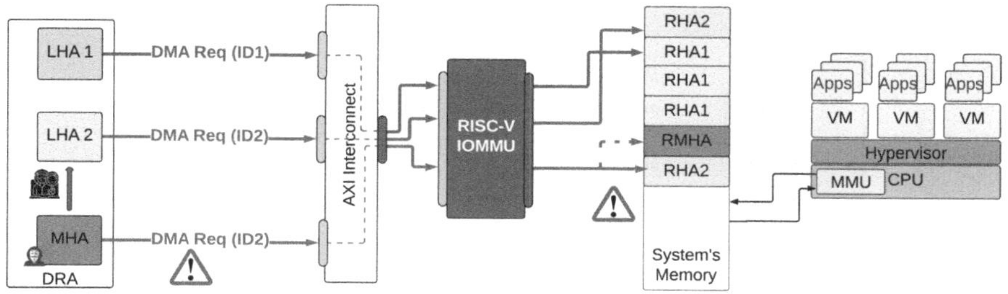

Fig. 1. Case study system architecture.

identity authentication within the IOMMU to subvert this isolation. Specifically, by modifying its Device ID at runtime, the MHA can impersonate a Legitimate Hardware Accelerator (LHA) and issue spoofed DMA translation requests that are erroneously authorized by the IOMMU, enabling unauthorized access to protected memory regions.

This exploitation stems from a fundamental assumption in the current RISC-V IOMMU design: the CPU, memory subsystem, interconnect, and IOMMU configuration form a trusted computing base, while peripherals - particularly dynamically reconfigurable HAs - are not authenticated beyond the Device ID they provide. In addition to identity-based privilege escalation, a MHA may also attempt to degrade system availability by generating high-rate request bursts, accumulating excessive outstanding transactions, or provoking interrupt storms, ultimately saturating shared hardware resources and affecting the execution of co-resident HAs. In this context, HAs in the DRA are therefore modeled as untrusted entities capable of issuing adversarial traffic, while the remaining system components are considered trusted.

3 Proposed Solution: ARMOR

The proposed ARMOR mechanism introduces a dedicated security wrapper positioned between HAs and the IOMMU, acting as an enforcement and monitoring layer for DMA transactions. Its primary objective is to guarantee the authenticity and non-forgeability of Device IDs associated with outgoing translation requests, thereby preventing privilege escalation through ID spoofing. Beyond enforcing identity integrity, the wrapper integrates lightweight observability and runtime monitoring to detect adversarial behaviors such as abnormal request rates, outstanding-transaction saturation, or interrupt flooding, and to react through controlled throttling and secure backpressure when necessary. These security enhancements are designed to incur a minimal resource footprint and negligible latency overhead, while fully preserving compatibility with dynamic partial reconfiguration and multi-HA coexistence in heterogeneous RISC-V systems.

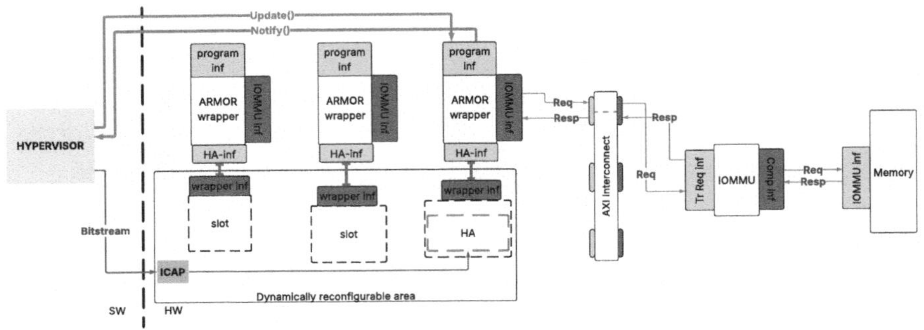

Fig. 2. Positioning of the ARMOR wrapper in SoC.

3.1 Architecture

To counter malicious behaviors originating from untrusted HA, we introduce ARMOR, a hardware-assisted security mechanism that combines identity enforcement with continuous runtime supervision. As shown in Fig. 2, ARMOR follows a wrapper-based architecture in which a dedicated security wrapper is instantiated for every slot within the DRA. Positioned between each slot and the IOMMU, the wrapper acts as a trusted intermediary responsible for validating and regulating all DMA translation requests and accesses.

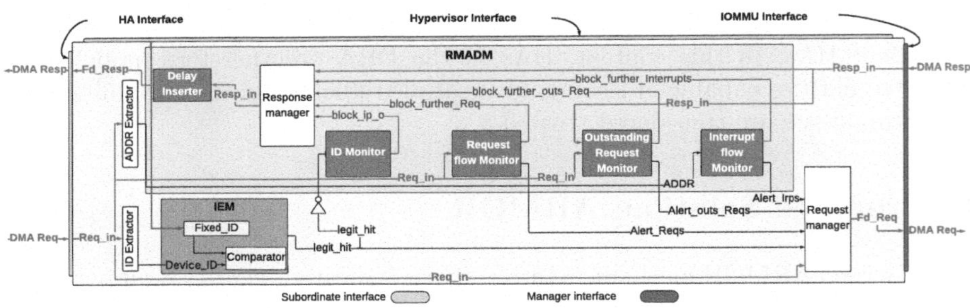

Fig. 3. ARMOR Wrapper's architecture.

At system initialization, the DRA is partitioned into predefined slots, each permanently associated with its own ARMOR wrapper. Every wrapper embeds a protected Device ID register, initialized to a neutral value (None) and updated dynamically based on the HA deployed in the corresponding slot. All authorized HAs are pregenerated offline, and their unique Device IDs are stored and managed by the hypervisor.

During runtime, when a software application requests the deployment of an HA, the hypervisor first programs the corresponding wrapper with the appropriate Device ID before loading the bitstream into the DRA. This guarantees that all subsequent transactions issued by the HA are automatically tagged with a trusted, hardware-assigned ID, thereby preventing any attempt at Device ID spoofing. Conversely, when an HA is removed or replaced, the wrapper's Device ID register is reset to None, ensuring that no stale or unauthorized requests can propagate. By maintaining a continuous one-to-one binding between each slot and its authenticated Device ID, ARMOR ensures that the IOMMU operates on reliable identity information, preserving memory isolation and eliminating ID forgery risks during dynamic reconfiguration.

As illustrated in Fig. 3, the wrapper exposes three AXI interfaces: two subordinate ports and one manager port. The first subordinate interface connects to the HA, allowing ARMOR to intercept and analyze all outgoing translation requests. The wrapper's manager interface forwards these requests to the IOMMU and operates in a transparent pass-through mode under normal, violation-free conditions with negligible added latency (2 cycles).

The second subordinate interface connects the wrapper to the hypervisor, which programs valid Device IDs at runtime and receives notifications of detected anomalies. This bidirectional channel ensures continuous consistency with the system configuration and enables privileged-level monitoring and enforcement. As ARMOR assures a gatekeeping action in an AXI-based system, request blocking—as used in the following sections— refers to suppressing a transaction by holding the wrapper's AxVALID low and simultaneously deasserting the device-facing AxREADY on the R/B channels, thereby halting propagation and generating an AXI SLVERR response to the HA.

Identity Enforcement Module (IEM): The ARMOR wrapper ensures that all transactions issued by a HA are associated with an authentic and non-forgeable Device ID, thereby eliminating opportunities for ID spoofing and unintended privilege escalation. This guarantee is enforced by the IEM, which performs strict in-line verification of transaction IDs before they reach the IOMMU.

For each outgoing request, the IEM extracts the Device ID transmitted on the AxMMUSID signal and compares it against a trusted reference securely stored within the wrapper and uniquely bound to the reconfigurable slot. This reference value, provisioned by the hypervisor during HA deployment, establishes a reliable and immutable association between slot and device identity.

Requests carrying a valid ID are forwarded transparently to the IOMMU. In contrast, any discrepancy triggers an immediate containment response: the wrapper blocks the request. This mechanism prevents any malformed or adversarial transaction from reaching the IOMMU, thereby reinforcing the integrity of the system's memory isolation policies.

Each failed verification is concurrently reported to the monitoring subsystem to support traceability and contribute to ARMOR's broader runtime anomaly-detection capabilities.

Runtime Monitoring and Anomaly Detection Module (RMADM). Complementing the IEM, ARMOR incorporates a RMADM responsible for supervising HA behavior throughout execution. This module continuously observes traffic patterns, transaction timing, and system-level interactions to detect malicious or anomalous activities that may threaten system availability or compromise data integrity. The monitoring logic operates along three complementary dimensions: identity traceability, Denial-of-Service (DoS) attacks mitigation, and timing-attack resilience.

Identity Traceability: The ID Monitor, records all verification outcomes reported by the IEM. By maintaining a history of mismatches and their frequency, it establishes identity traceability over time, enabling the detection of repeated or systematic spoofing attempts. When the number of ID mismatches exceeds a programmable threshold, the monitor asserts a suspicious activity flag, notifying the hypervisor. This allows higher-level software to initiate adaptive countermeasures such as slot isolation, bitstream revalidation, or partial reconfiguration. This mechanism provides a lightweight but effective form of accountability, ensuring that every reconfigurable slot maintains a verifiable behavioral history with respect to identity integrity.

Denial of Service Attacks Mitigation: To preserve system availability, the RMADM includes several traffic regulation mechanisms designed to detect and mitigate *DoS* originating from compromised HAs. It specifically addresses three major forms of resource exhaustion described below.

(1) A Request *Storm Attack* described in [16] arises when a MHA issues DMA translation requests at an excessively high rate, saturating the interconnect and the IOMMU queues, thereby delaying or blocking legitimate traffic and causing system-wide degradation or DoS. To detect and mitigate such attacks, the ARMOR wrapper embeds a transaction-rate monitor that measures the frequency of outgoing requests over a sliding time window using a lightweight counter. When the request count exceeds a programmable threshold, the wrapper flags a rate anomaly. Upon detecting a storm, the wrapper enforces a windowed throttling policy by modulating the output handshake. If the request rate exceeds 50% of the allowed per-window quota, the wrapper switches to a 1:1 admission policy—forwarding one request and blocking the next. When the maximum per-window allowance is reached, the wrapper enters a temporary blocking state, rejecting all remaining transfers until the window resets. This adaptive backoff suppresses the offending HA's throughput while preserving service for benign components. Persistent violations across successive windows trigger escalation to the hypervisor and event logging for forensic analysis. All thresholds and window parameters are configurable to balance detection sensitivity and false-positive control.

(2) The *Outstanding Requests Overflow* attacks as discussed in [16] occur when a MHA floods the interconnect or IOMMU with transactions without waiting for their responses, saturating outstanding request buffers and starving LHA. To detect and mitigate such attacks, the ARMOR wrapper includes an outstanding-request counter that increments on every issued request (AxVALID asserted) and decrements upon completion (R/BVALID). If the count surpasses

a configurable threshold, and the condition persists across successive monitoring windows the wrapper reports an overflow anomaly. When an overflow is detected, the wrapper immediately applies backpressure by blocking new requests to the IOMMU until all pending transactions complete and the counter returns to zero. If overflows recur across multiple windows, the event is escalated to the hypervisor. This prevents resource exhaustion while preserving fairness and system availability.

Message-Signaled Interrupts (MSIs) are widely used in heterogeneous SoCs: an HA signals an event by issuing a memory-mapped write to a predefined MSI address, which the interrupt controller interprets as an interrupt request. Although efficient, each interrupt imposes a substantial CPU cost—pipeline flush, context save/restore, cache perturbation, and execution of a privileged ISR. At high frequency, these overheads accumulate rapidly and can dominate CPU time.

(3) An *Interrupt Storm Attack* arises when a compromised HA issues interrupts at an abnormally high or bursty rate, overwhelming the CPU or IOMMU interrupt-handling subsystem and delaying legitimate events. To detect and mitigate such attacks, the ARMOR wrapper detects interrupt-related transactions by inspecting AXI writes to the MSI address (AxADDR) and monitors their temporal density through a dedicated interrupt-rate counter evaluated over successive time windows. Two abnormal patterns are distinguished: (i)Continuous storms: sustained high-frequency interrupt issuance exceeding a programmable threshold and (ii) Discontinuous storms: repetitive shorter bursts identified through an abnormal interrupt-to-DMA ratio. Upon detecting an interrupt-rate violation, ARMOR initiates a temporary interrupt blocking phase, suppressing MSI writes from the offending slot while allowing DMA traffic to continue under controlled admission. If anomalous behavior persists across multiple windows, the wrapper escalates to full interrupt masking for that slot and signals the hypervisor for higher-level remediation (e.g., slot suspension, reconfiguration, isolation). By enforcing localized interrupt control, ARMOR prevents CPU overload and preserves deterministic system behavior even under malicious or unstable HA activity.

Timing Attack Mitigation: *Timing-based side channels* [3] exploit subtle latency variations—such as response delays, arbitration timing, or IOTLB access latency—to infer information about protected memory regions or concurrent device activity. Although such attacks do not directly compromise integrity, they can enable cross-tenant information leakage in shared hardware environments. To reduce this leakage risk, the ARMOR wrapper implements a preventive countermeasure based on controlled jitter insertion. Pseudo-random timing offsets are introduced into specific protocol responses thereby perturbing the precise timing information observable by a potential adversary. The injected jitter obscures fine-grained latency patterns without altering functional behavior or protocol compliance. This mechanism effectively lowers the temporal resolution available to attackers, making statistical timing inference unreliable while imposing negligible performance overhead. By integrating this lightweight obfuscation directly into the data path, ARMOR proactively mitigates timing-based side channels without requiring complex detection or monitoring logic.

3.2 Evaluation

In the following, we describe only how ARMOR mitigates the ID spoofing attacks discussed in [5]. ARMOR was evaluated in a RISC-V-based environment featuring two I/O peripherals (HA) interacting with the IOMMU. In the present prototype, the hardware implementation focuses on the identity verification and ID monitoring module that govern transaction-level authentication. Each HA is assigned a unique Device ID, securely stored within its ARMOR wrapper, and all translation requests pass through this wrapper before reaching the IOMMU. The system relies on the RISC-V IOMMU implementation from [14], compliant with specification v1.0.

Functional simulations were performed using Xilinx Vivado v2022.1 to analyze ARMOR's behavior under ID spoofing. In this scenario, a MHA attempts to impersonate another HA by issuing write requests while repeatedly modifying its Device ID. As shown in Fig. 4, the wrapper intercepts each request, extracts the transmitted ID (*Device_ID_o*), and compares it to the authorized ID provisioned by the hypervisor (*fixed_ID_reg*). Requests carrying forged IDs (*legit_hit* = 0) are blocked (*req_wrapper_iommu_o.aw_valid* = 0), while only the request with the legitimate ID is forwarded. The ID monitor relies on an 8-bit counter, allowing the hypervisor to configure a threshold, which is set to three in this example. After three consecutive mismatches (*failure_count* = 3), the wrapper raises a threat alert (*threat_detected* = 1) and halts all subsequent transactions from the compromised HA (*block_ip_o* = 1) until re-authorization.

These results confirm that ARMOR enforces device identity authenticity and prevents unauthorized DMAs while leveraging the IOMMU as the central memory management and access-control component.

The ARMOR wrapper achieves pure hardware-level privilege enforcement with minimal integration cost, providing a scalable and low-latency alternative to more complex, software-assisted or memory-dependent security mechanisms. Indeed, the IEM relies solely on a comparator, while the RMADM is primarily composed of counters. No memory arrays, FIFOs, or large control structures are required, which drastically reduces both area and power consumption. This simplicity not only minimizes the critical path and timing impact but also enhances portability and reusability across different FPGA or ASIC platforms. To evaluate the hardware overhead, the ARMOR wrapper was synthesized as an identity verification module on a Xilinx UltraScale+ FPGA board. The synthesis results therefore reflect the hardware cost of the IEM and the ID monitor, which protect the system against the attacks described in [5] while the remaining countermeasures—although part of the overall architecture—are not part of the current synthesis scope. The design exhibits minimal resource usage—129 LUTs (0.06%) and 500 registers (0.11%), with no BRAM or DSP consumption—demonstrating strong identity protection with negligible area cost, suitable for integration into reconfigurable heterogeneous systems.

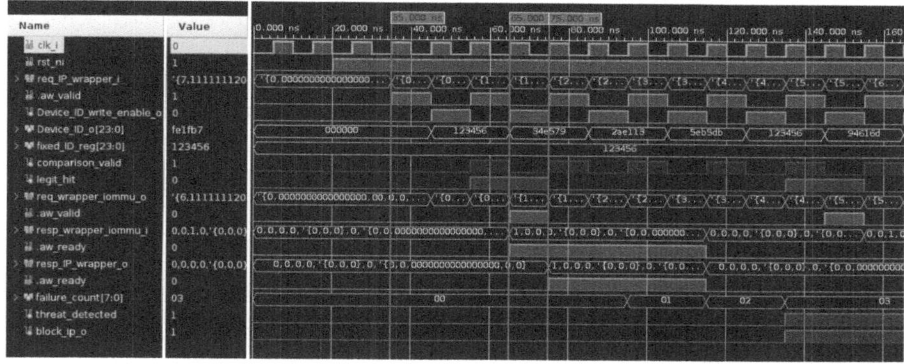

Fig. 4. Waveform analysis of Device ID verification through ARMOR wrapper.

4 Discussion

Securing HAs and DMA transactions in heterogeneous, reconfigurable SoCs has been widely studied. Hardware-based solutions provide low-latency isolation but often lack dynamic identity verification and continuous supervision. Software- and hypervisor-based approaches offer flexibility and observability at the cost of higher overhead, while AXI interconnect monitoring frameworks improve anomaly detection yet rarely enforce device identity. In the following sections, we review these approaches and show how ARMOR unifies identity verification and real-time supervision in a lightweight framework.

Hardware-Based Protection Mechanism: Several hardware-based solutions have been proposed to secure HAs and DMA transactions in heterogeneous and reconfigurable SoCs. RtrustSoC [8] offers a comprehensive security architecture with reconfigurable regions and a secure-by-design approach. The architecture employs a multi-world segregation model that extends the trust concept to hardware IPs. All system components are encapsulated within distributed, hardware-coded security wrappers that collectively form the architectural backbone. These wrappers continuously monitor SoC bus transactions and enforce dynamic isolation by validating IP and world IDs against defined security policies to prevent unauthorized access. TrustToken [1] targets FPGA-accelerated cloud platforms by generating cryptographic identity tokens at runtime via a hybrid ring-oscillator PUF, removing the need for non-volatile key storage. Tokens are verified by a central controller to authorize third-party IP interactions, ensuring secure execution in multi-tenant environments. Similarly, SAM [4] protects multi-tenant FPGA systems from address-redirection and task-hiding attacks by continuously authenticating active hardware and software entities using challenge–response protocols with Message Authentication Codes, ensuring secure execution.

FPGA shells adopt a different approach by enforcing hardware-level isolation and controlled memory management. Coyote [6], for example, uses a hierarchical architecture to encapsulate FPGA applications as vFPGAs, each with its own address space, and integrates TLBs into the wrappers. Unauthorized DMA attempts trigger TLB misses or page faults, emulating IOMMU behavior.

Other solutions target DMA protection in reconfigurable systems but are often limited in adaptability or scope. Optimus [7], a hypervisor-based solution, isolates virtual accelerator DMAs through page table slicing on shared-memory FPGAs, but its static partitioning restricts integration with dynamic reconfiguration. MeetGO [10], TREE [11], and ShEF [15] leverage reconfigurability to create trusted execution environments: MeetGO isolates applications via physical partitioning and secure agents, TREE uses DPR to instantiate trusted tiles with dedicated memory, and ShEF secures accelerator-to-memory communication via authenticated encryption. While effective for application isolation or data confidentiality, these frameworks prioritize application isolation or data confidentiality rather than comprehensive control over DMA behaviors during reconfiguration.

Runtime Monitoring and Supervision Within AXI Interconnects: A growing body of research has addressed runtime monitoring in AXI-based interconnects to improve the predictability and security of heterogeneous SoCs. [2] introduces a modular extension that regulates bandwidth through credit-based control and transfer fragmentation, thereby achieving temporal isolation among mixed-criticality components. Building on similar principles, [12] integrates bandwidth reservation mechanisms at the hypervisor level to ensure predictable communication between HAs and shared resources. At the security layer, [13] employs access control wrappers to intercept and block unauthorized transactions at their origin, preventing interference and DoS conditions. In addition to these works, tools such as [16] enhances protocol assurance by analyzing the security of the interconnect and modeling the functional and security semantics of the AXI standard, enabling early detection of vulnerabilities caused by design inconsistencies. [8] also features a dynamic reconfigurable penalty system which monitors malicious third-party IPs for abnormal behavior and can temporarily disable their communication ability, thus mitigating some DoS attacks.

Positioning of ARMOR: The proposed ARMOR mechanism positions itself as a lightweight yet effective enhancement to existing SoC protection flows by extending security guarantees to dynamically reconfigurable HAs without disturbing the strong baseline already provided by the IOMMU for static system components. A key differentiating aspect is ARMOR's tight integration with the IOMMU: rather than adding parallel translation units, static shells, or distributed access-control tables, ARMOR reinforces the IOMMU's role as the central and sole authority for memory isolation by ensuring that all requests entering the IOMMU are annotated with trusted Device IDs and have been behaviorally validated at runtime. This synergy preserves a single, globally consistent translation domain while maintaining end-to-end protection across reconfiguration events, preventing state drift or stale identity bindings during dynamic updates.

Compared to FPGA shells, ARMOR introduces only minimal hardware logic and no additional TLBs or address-translation layers. Compared to architectures like RtrustSoC, whose overhead grows with the number of worlds, access-rights tables, and wrapper components, or TrustToken and SAM, which require PUF primitives, key-storage logic, authentication controllers, and cryptographic challenge–response flows, ARMOR provides identity enforcement and runtime supervision with significantly lower complexity. Its unified, IOMMU-centric approach thus offers a fine-grained alternative to coarse-grained shells and a lighter, more practical solution than cryptographic identity frameworks, while remaining easily adaptable to heterogeneous and reconfigurable SoCs.

5 Conclusion and Future Work

This paper presented ARMOR, a lightweight hardware-assisted mechanism that enforces device identity integrity and provides runtime supervision of dynamically reconfigurable accelerators, while reinforcing the IOMMU as the central authority for memory isolation. ARMOR ensures that all DMA transactions carry authenticated IDs and that anomalous behaviors are detected and mitigated with minimal latency and hardware overhead, preserving protection continuity across reconfiguration events. As future work, we aim to evaluate ARMOR under a broader range of threat scenarios beyond ID spoofing, in order to assess the effectiveness of its remaining architectural countermeasures. We also plan to extend ARMOR with a hypervisor-level software service that analyzes notifications from the wrapper to detect more sophisticated attacks, such as memory scans or probing. This service will implement adaptive policies and intelligence while maintaining minimal overhead and will be tightly integrated with the reconfiguration management system, providing a unified hardware–software security framework for dynamic heterogeneous SoCs.

Acknowledgement. This work was carried out within the TrustGW project (ANR-21-CE39-0005), funded by the French National Research Agency (ANR).

References

1. Ahmed, M.K., Saha, S.K., Bobda, C.: Trusted IP solution in multi-tenant cloud FPGA platform. In: 2022 IEEE 8th World Forum on Internet of Things, pp. 1–6 (2022)
2. Benz, T., et al.: AXI-REALM: safe, modular and lightweight traffic monitoring and regulation for heterogeneous mixed-criticality systems. IEEE Trans. Comput. **74**(9), 3072–3086 (2025)
3. Bossuet, L., Benhani, E.M.: Security assessment of heterogeneous soc-FPGA: on the practicality of cache timing attacks. In: 2021 IFIP/IEEE 29th International Conference on Very Large Scale Integration, pp. 1–6 (2021)
4. Elnaggar, R., Karri, R., Chakrabarty, K.: Multi-tenant FPGA-based reconfigurable systems: attacks and defenses. In: 2019 Design, Automation & Test in Europe Conference & Exhibition, pp. 7–12 (2019)

5. Jendoubi, A., Prévotet, J.C., Tanguy, P., Cotret, P.: Security of dynamically reconfigurable RISC-V systems: I/O attack focus. In: 2025 IEEE International Parallel and Distributed Processing Symposium Workshops, pp. 1220–1220 (2025). https://doi.org/10.1109/IPDPSW66978.2025.00198
6. Korolija, D., Roscoe, T., Alonso, G.: Do OS abstractions make sense on FPGAs? In: 14th USENIX Symposium on Operating Systems Design and Implementation, pp. 991–1010 (2020)
7. Ma, J., et al.: A hypervisor for shared-memory FPGA platforms. In: Proceedings of the Twenty-Fifth International Conference on Architectural Support for Programming Languages and Operating Systems, pp. 827–844 (2020)
8. Milan, R., Bossuet, L., Lagadec, L., Andres Lara-Nino, C., Colombier, B., Bollengier, T.: Efficient adaptive multi-level privilege partitioning with RTrustSoC. IEEE Trans. Circuits Syst. I Regul. Pap. **72**(2), 497–509 (2025)
9. Morgan, B., Alata, E., Nicomette, V., Kaaniche, M.: IOMMU protection against I/O attacks: a vulnerability and a proof-of-concept. J. Braz. Comput. Soc. **24** (12 2018)
10. Oh, H., Nam, K., Jeon, S., Cho, Y., Paek, Y.: MeetGo: a trusted execution environment for remote applications on FPGA. IEEE Access **9**, 51313–51324 (2021)
11. Pereira, S., Gomes, T., Cabral, J., Pinto, S.: TREE: bridging the gap between reconfigurable computing and secure execution. IACR Trans. Cryptographic Hardware Embedded Syst. **2025**(3), 115–138 (2025)
12. Restuccia, F., Biondi, A., Marinoni, M., Cicero, G., Buttazzo, G.: AXI Hyperconnect: a predictable, hypervisor-level interconnect for hardware accelerators in FPGA SoC. In: 2020 57th ACM/IEEE Design Automation Conference, pp. 1–6 (2020)
13. Restuccia, F., Meza, A., Kastner, R.: AKER: a design and verification framework for safe and secure soc access control. In: 2021 IEEE/ACM International Conference on Computer Aided Design, pp. 1–9 (2021)
14. Zero Day Labs: RISC-V IOMMU (2025). https://github.com/zero-day-labs/riscv-iommu. Accessed 12 Mar 2025
15. Zhao, M., Gao, M., Kozyrakis, C.: ShEF: shielded enclaves for cloud FPGAs. In: Proceedings of the 27th International Conference on Architectural Support for Programming Languages and Operating Systems, pp. 1070–1085. ASPLOS '22, Association for Computing Machinery, New York, NY, USA (2022)
16. Zonta, M., Hinderling, N., Shinde, S.: Xray: detecting and exploiting vulnerabilities in arm AXI interconnects. In: 2025 Design, Automation & Test in Europe Conference, pp. 1–7 (2025)

BERT-Inspired HT Localization on FPGA AI Accelerators

AbdelSalam Baltagi(✉), Yehya Nasser, and Amer Baghdadi

IMT Atlantique, Lab-STICC, UMR CNRS 6285, F-29238 Brest, France
abdel-salam.baltagi@imt-atlantique.fr

Abstract. Outsourcing accelerates AI hardware development but exposes designs to hardware Trojan (HT) risks. The stealthy nature of HTs and increasing complexity of CNN accelerators limit the effectiveness of traditional detection and localization techniques. This paper presents a segmentation-driven power side-channel framework combined with a customized BERT-inspired model to localize HT-affected regions relative to a trusted baseline. By dividing traces into segments and analyzing them with transformer attention, the method enables targeted auditing and reduces manual inspection. Experiments on real power measurements with physical HTs achieve 92.1% localization accuracy on known HTs and 75.2% on unseen HTs. Contributions include a preprocessing pipeline for segmented trace analysis, a BERT-style architecture optimized for HT localization, and demonstrated generalization across multiple HT instances. The approach supports efficient, focused security audits in outsourced fabrication workflows.

Keywords: Hardware Trojan · AI Accelerator · Machine learning · BERT-inspired models · Power side-channel analysis

1 Introduction

As HW performance and algorithms advance, the rapid growth of AI, particularly neural networks, has shifted from research to industrial deployment [7]. Modern ICs commonly integrate third-party IP, especially AI accelerators, but outsourcing creates security risks, most notably hardware Trojans (HTs). Detecting and localizing these threats has therefore become essential.

CNN-based accelerators are widely used in critical applications such as image classification, face recognition, object detection, and medical diagnosis [14]. Their large size, massive parallelism, and high resource utilization make them fundamentally more complex than classical designs [1], complicating HT detection and increasing the impact of fabrication-time tampering.

Existing HT detection techniques suffer from limited coverage, precision, and scalability, and most focus solely on detection rather than localization [20]. For modern AI accelerators, identifying only the presence of an HT leaves auditors with the impractical task of manually inspecting thousands of blocks. Given

M. Kowalczyk and C. Monière (Eds.): DASIP 2026, LNCS 16512, pp. 159–170, 2026.
https://doi.org/10.1007/978-3-032-23871-9_13

rising deployment demands, lengthy testing cycles are costly, underscoring the need for fast and accurate localization.

Localizing HT-impacted regions therefore becomes crucial [13]. Building on our prior work that successfully detected HTs in AI accelerators [2], we now address localization. While detection flags malicious modifications, localization identifies the affected blocks, enabling region-level threat assessment instead of impractical full-circuit review.

HT detection and localization approaches are divided into side-channel analysis (SCA) and netlist-based testing [4]. SCA examines physical signals such as power, timing, or electromagnetic emissions, while netlist methods analyze structural anomalies. However, netlist-based scalability is questionable for large accelerators, and third-party workflows may not provide full design access. In contrast, power SCA (PSCA) reflects the behavior of deployed HW [18].

ML-assisted PSCA has shown promise in cryptographic circuits [3,28], but commonly used ML models struggle with the long sequences of AI accelerators. Truncating traces loses essential context, and sequence models like Long Short-Term Memory (LSTM) cannot reliably process very long inputs. Because small deviations across sub-traces may collectively reveal HT activity, traces must be analyzed holistically rather than in isolated segments [8].

Recent advances in Transformer architectures show their ability to capture long-range dependencies in sequential data. This capability, absent in traditional PSCA techniques, enables full-trace contextual analysis, providing deeper insight into HW behavior. To the best of our knowledge, this is the first work to apply language-model concepts to localize HT-impacted regions in AI accelerators using PSCA, rather than only detecting HTs in conventional circuits.

This work targets HT-impacted regions localization in AI accelerators using Bidirectional Encoder Representations from Transformers (BERT) with PSCA. Our main contributions are:

1. **Novel formulation:** We are the first to reformulate power trace analysis as a language-modeling task, treating segmented sub-traces as words and the entire trace as a sentence.
2. **Context-aware localization:** A transformer-based model captures contextual dependencies across parallel circuit blocks, enabling anomaly disentanglement under heavy parallelism.
3. **Dual-task training:** Joint feature-recovery and anomaly-scoring tasks improve localization accuracy and generalization to unseen HTs.

The rest of the paper is organized as follows: Sect. 2 reviews related work; Sect. 3 describes the methodology; Sect. 4 presents experimental results; Sect. 5 concludes the paper.

2 Related Work

This section reviews prior work on HT localization, highlighting different methodologies. Also, we discuss advances in language modeling techniques, as they provide the foundation for our BERT-inspired approach to HT localization.

2.1 Hardware Trojan Localization

Table 1 summarizes recent HT localization efforts. Netlist information is the predominant feature source; only one work uses path-delay data. Both [20,29] convert the netlist into a graph and apply Graphical Neural Network (GNN) to identify HT locations relative to a clean reference. Other studies adopt graph-based representations using Graph Attention Network [27], Signal Transfer Graph [6], or Graph Convolutional Network (GCN) before the Transformer [13], with evaluations primarily on cryptography or communication benchmarks.

In [26], the netlist is treated as text and analyzed with a text CNN for localization, while [9] employs a text-based, word-statistics approach to identify HT-vulnerable macro-blocks, tested on Digital Signal Processing (DSP) IPs. A different direction is [19], which uses path-delay information, forms a graph, and applies a GCN, evaluated on International Symposium on Computer Architecture (ISCA) and International Test Conference (ITC) benchmarks.

These techniques report 85–95% localization accuracy but remain limited to classical, smaller benchmarks with short execution times. Their scalability to AI accelerators is unclear, and their reliance on netlist-derived information is problematic since design access may not be available in outsourced flows. This highlights a gap for HT localization using PSCA for AI accelerators.

Table 1. Comparison of recent HT localization methods

Paper	Source	Technique	Application
[20]	Netlist Analysis	Graph Neural Network	Crypto/Comm. Algorithms
[27]	Netlist Analysis	Graph Attention Network	Crypto/Comm. Algorithms
[6]	Netlist Analysis	Signal Transfer Graph	Crypto/Comm. Algorithms
[13]	Netlist Analysis	GCN + Transformer	Crypto/Comm. Algorithms
[29]	Netlist Analysis	Graph Neural Network	Crypto/Comm. Algorithms
[26]	Netlist Analysis	Text CNN	Crypto Algorithms
[19]	Path-delay Analysis	GCN	ISCA & ITC algorithms
[9]	Netlist Analysis	Word Statistics	DSP IPs

2.2 Language Modeling Techniques

HT localization in power traces can be framed as time-series anomaly detection using prediction-based [15–17] or reconstruction-based methods [11,21]. While effective on short, periodic signals, these methods struggle with the long, highly aperiodic power consumption traces of AI accelerators. Our preliminary tests confirm these limitations, motivating the use of language-modeling techniques to capture long-range dependencies.

Language modeling evolved from N-grams and Word2Vec to Recurrent Neural Network (RNN) and LSTM, which improved tasks such as translation and classification [12]. Attention mechanisms then enabled Transformer architectures [24], with BERT [5] as a widely adopted encoder-only model. BERT captures contextual meaning, e.g., distinguishing river bank vs. bank robber, supports long sequences, and provides rich token- and sentence-level representations.

BERT employs only the Transformer encoder. Tokens are mapped to d_{model}-dimensional embeddings, combined with positional and segment embeddings, then processed through stacked encoder layers. Pre-training uses Masked Language Modeling (MLM) and Next Sentence Prediction (NSP), producing a contextual encoder that can be paired with decoders or lightweight classifiers (e.g., MLPs) for downstream prediction.

3 Methodology

This section outlines our methodology, shown in Fig. 1, consisting of five main steps: experimental setup and ML model implementation on the FPGA, power measurement and HT benchmark preparation, data pre-processing, and finally the development of the BERT-inspired localizer.

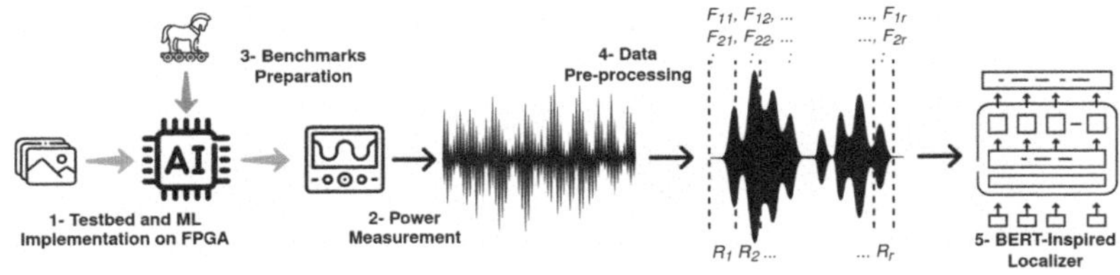

Fig. 1. Main steps of the proposed methodology.

3.1 ML Implementation on FPGA

Our testbed shown in Fig. 2 uses an Artix-7 (xc7a100tftg256-2) FPGA, serving as a representative AI accelerator. The setup employs the ChipWhisperer-305 (CW305) board from NEWAE, which includes an ATMEL microcontroller for USB communication and provides an external trigger (TRIG) to synchronize FPGA operations with measurement equipment such as an oscilloscope.

For the implemented AI model, we used the FINN framework [23] to generate HDL for the accelerator. Instead of training a model from scratch, we adopted the pre-trained quantized CNV CNN from the Brevitas library, fully compatible with FINN. The CNV architecture includes six convolution layers and three fully connected (FC) layers; we added input normalization and a label-selection layer for output class retrieval. The FINN compiler then produced the architecture in Verilog, and Vivado was used to integrate communication blocks for image transfer and prediction extraction between the host and FPGA.

The `ThresholdingBatch` block performs input normalization. The `Matrix VectorActivation` (MVAC) block implements FC layers and optional activation, and also handles convolution operations after the `ConvolutionInput Generator`, which reformats inputs for matrix multiplication. Max-pooling is mapped to `StreamingMaxpool`, dropout has no HW mapping, and the TopK functionality is mapped to the `LabelSelector`. Additional intermediate blocks ensure proper dataflow integration.

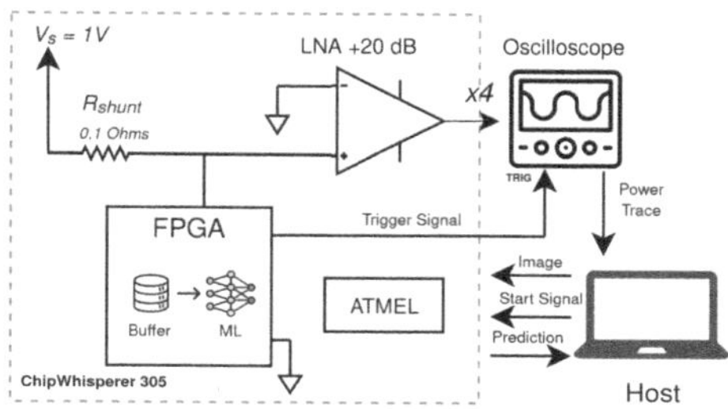

Fig. 2. Schematic of the testbed.

For the CNV model, the final design contains 19 main blocks: `Thresholding Batch` ×1, `ConvolutionInput- Generator` ×6, `MVAC` ×9, `StreamingMaxpool` ×2, and `LabelSelector` ×1.

3.2 Power Measurement

As shown in Fig. 2, the CW305 target board includes a 0.1Ω shunt resistor placed in series between the $V_s = 1$V supply and the FPGA. The voltage drop across the shunt is routed into a low-noise amplifier (LNA), which provides 20 dB of gain at pin x4 and removes the DC component. A Teledyne LeCroy HDO9409-MS oscilloscope (10-bit ADC) records the amplified signal. The oscilloscope is triggered on the rising edge of the FPGA's TRIG signal to ensure precise alignment.

The FPGA runs at 20 MHz, and the oscilloscope samples at 100 MHz, following the 5× rule recommended in [22]. Traces are cropped to retain only the inference window. For each benchmark, measurements are obtained using a predefined subset of CIFAR-10 images. All captured traces are transferred to the host and stored with metadata using a Python script enabling rapid and consistent acquisition of large trace datasets.

3.3 Benchmarks Preparation

The benchmarks are divided into two categories: known HTs for training, validation, and testing, and unknown HTs for external evaluation. All HTs were inserted using incremental implementation, with the clean CNV design as the reference checkpoint to minimize insertion impact, resulting in an additional HW overhead of only ∼0.0015%. Note that the HTs are designed not to affect block timing; otherwise, simple timing analysis would suffice.

Known HT Benchmarks. Known HTs were designed to be minimal to enable high-resolution localization. A small dummy HT was iteratively inserted into each layer to emulate realistic Trojan behavior while allowing the model to learn localized anomalies. This supports generalization to any HT mechanism, as shown later in Sect. 4.

The HT consists of four shift-register variants, shift-left with inverted Most Significant Bit (MSB), rotate-left, shift-right with inverted Least Significant Bit (LSB), and half-word swap, operating sequentially under an active-HIGH enable signal. The enable is asserted only when the corresponding CNV signal's `data_in_valid` or `data_out_valid` is HIGH, ensuring activation only during layer operation.

This HT was inserted into each of the 19 blocks described in Sect. 3.1, plus the clean version, producing 20 CNV designs (1 clean + 19 infected). The dataset was balanced at the region-label level and split into 4620 training, 990 validation, and 990 test samples.

Unknown HT Benchmarks. A total of 15 unknown HT benchmarks were implemented, each inspired by recent HT designs targeting AI accelerators. These benchmarks cover diverse trigger and payload mechanisms to provide a realistic and challenging evaluation set.

The first two Trojans (ext_comb_2 and ext_seq_4) employ external counters to detect specific pixel patterns or prediction sequences before forcing a malicious class output via a multiplexer at the output. The relu_fc7 Trojan uses an internal XOR-based trigger within the seventh `MVAC` block, modifying the ReLU output when activated.

Five Trojans adopt confidence-reduction strategies, including shifting classification scores (ht_shft_{10,5,1}_MsCls) or inducing targeted or swapped-class misclassification (ht_MxTgMs, ht_SwSTMs), all triggered by evaluating a slice of the `LabelSelector` input.

Three maxpool-based Trojans (maxpool_0_always-on, maxpool_0_pattern-comparator, maxpool_0_counter) modify the first `StreamingMaxpool` block by inverting its minimum/maximum decision under always-on, pattern-based, or sequential triggers.

The reconfigurable_mvac_0 Trojan flips the output of the first `MVAC` block upon matching a predefined vector, while the geometric_trojan activates when a specific image pattern is detected, applying a mask to the input.

Finally, two clock-glitch Trojans (clock_glitch_always-on and clock_glitch_pattern-recognition) manipulate the clock signal within the `LabelSelector` block, either continuously or based on input pixel patterns.

3.4 Data Pre-processing

To represent a long, highly fluctuating power trace as a sentence of words, we preprocess the data by treating the trace as a sequence of regions, where each region corresponds to a word. Each region is assigned a modification score that reflects resource utilization changes relative to the clean accelerator, and the model learns to predict these scores from region-level features.

Resource utilization is extracted per block from Vivado reports and compared against the clean design to compute the per-block modification score. Since multiple blocks may be active simultaneously, directly mapping activity to the power trace is ambiguous. To resolve this, we segment the trace according to the block activity timeline: whenever a block starts or stops, a new region is created, ensuring each region corresponds to a unique set of active blocks. This allows us to detect regions with abnormal modification scores and trace them back to the common active block(s), thereby identifying the infected component. In our case we had $r = 37$ consecutive regions.

Because regions vary in length, each is transformed into a compact feature vector. Four statistical features (MAD, skewness, PSD standard deviation, and spectral entropy) are extracted following [2], and PCA is applied for dimensionality reduction. Using $N_{\text{PCA}} = 20$ yields feature vectors of size $f = 24$. PCA is fitted on the training set only, and regions are zero-padded to the maximum length, producing sample matrix of dimension $(r, f) = (37, 24)$.

Region labels are obtained by summing the modification scores of the active blocks in each region and binarizing them using a threshold: 0 if lower than the threshold and 1 otherwise, producing label vector of dimension $(r) = (37)$.

3.5 BERT-Inspired Localizer

Each HT insertion modulates the entire power trace in a unique way, even under incremental implementations [2]. For this reason, we opt to benefit from the strong contextual modeling capabilities of transformers, especially BERT, as demonstrated in language understanding applications. As shown in Fig. 3, the model processes one sentence at a time (i.e., one power trace). Similar to standard BERT, each region feature vector is projected into an embedding of dimension d_{model}, producing a tensor of shape (r, d_{model}). To incorporate temporal context, relative positional embeddings are added to preserve the sequential order of regions, following [25], which is critical for power traces. The resulting embeddings are then passed through Transformer encoder layers to obtain contextualized hidden states of shape (r, d_{model}).

Based on empirical evaluation, we use 2 encoder layers, 4 attention heads for the Transformer encoder, and $d_{\text{model}} = 128$ for the model hidden dimension. At the output, two heads are attached to perform joint learning: (1) a linear head that predicts masked features, producing an output of shape (r, f) in an MLM-like task, and (2) an MLP head that predicts the binary modification score for each region, yielding a vector of shape (r).

This mechanism supports varying sequence length, by padding traces to the maximum length with attention masks ignoring padding for the Transformer, while the MLP naturally supports any value of r by iterating over regions.

During inference, only the score prediction vector is significant to indicate whether each region is impacted or normal, with all-normal regions signifying a clean accelerator. This work introduces two key novelties: framing power trace

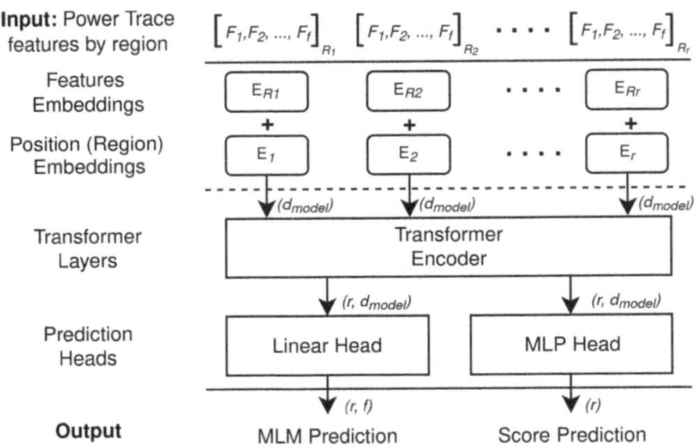

Fig. 3. Flow of the proposed BERT-inspired model.

analysis as a language-modeling task for HT localization, and leveraging dual-task training, feature recovery and anomaly scoring, to learn contextual patterns that generalize to unseen Trojans. The formulation is also scalable to variable-length traces through Transformer padding and iterative MLP processing.

4 Results and Discussion

We evaluate the proposed BERT-inspired localizer on known (in-distribution) and unknown (out-of-distribution) HT benchmarks. Accuracy is reported at the region level, and the threshold ($thresh$) was set to 2 based on experimental trials that yielded the best accuracy.

4.1 Known HTs

Table 2 reports the results on the test dataset, which is in-distribution with the training and validation data but still contains unseen samples. When training the model solely on the modification score prediction (score head only), we obtained a classification accuracy of 84.6%. In contrast, equipping the model with both the score head and the MLM head led to a significant improvement, yielding 92.1% classification accuracy and an R^2 score of 0.96 for the regression task. This boost clearly illustrates the benefit of the MLM objective: by forcing the model to reconstruct masked features, it more effectively captures contextual dependencies across regions, thereby improving anomaly localization accuracy.

4.2 Unknown HTs

Table 3 shows per-HT results for the proposed model with the MLM head across thresholds ($thresh = 1, 2, 3$), with average accuracies of 70.5%, 75.2%, and

Table 2. Test dataset results with and without MLM task (*thresh* = 2)

Test Data	**Accuracy**	R^2 **Score**
Score head only	84.6%	–
MLM + Score heads	92.1%	0.96

74.9%. The lower accuracy at *thresh* = 1 shows increased difficulty distinguishing noise from anomalies, so we adopt *thresh* = 2 for the following discussion, yielding the highest average accuracy.

Table 3. Per-HT accuracy on the evaluation dataset across threshold values (*thresh* = 1, 2, 3) using the MLM head

HT Name	***thresh* = 1**	***thresh* = 2**	***thresh* = 3**
ext_comb_2	41.8%	36.2%	43.7%
ext_seq_4	60.4%	60.2%	59.4%
relu_fc7	87.8%	91.8%	85.8%
ht_shft_10_MsCls	89.0%	94.4%	91.4%
ht_shft_5_MsCls	72.5%	72.9%	71.0%
ht_shft_1_MsCls	71.2%	72.0%	70.1%
ht_MxTgMs	91.9%	94.0%	86.4%
ht_SwSTMs	98.2%	99.7%	92.1%
maxpool_0_always-on	65.9%	65.0%	65.2%
maxpool_0_pattern-comparator	66.6%	65.0%	64.8%
maxpool_0_pattern-counter	59.3%	57.0%	57.0%
reconfigurable_mvac_0	59.3%	55.4%	61.5%
geometric_trojan	70.6%	72.8%	70.3%
clock_glitch_always-on	88.7%	98.8%	95.6%
clock_glitch_pattern-recognition	87.3%	92.9%	97.0%
Average	70.5%	75.2%	74.9%

Our observations show a strong correlation between HT detection accuracy and the parallelization level of FPGA regions. HTs in less parallelized layers achieve higher accuracy, as fewer active HW blocks reduce noise and improve the signal-to-noise ratio, making subtle anomalies easier to detect. Consistently, benchmarks including relu_fc7, ht_shft_10_MsCls, ht_MxTgMs, ht_SwSTMs, clock_glitch_always-on, and clock_glitch_pattern-recognition reached high localization accuracy (91–99.7%) in these low-parallelization regions.

Conversely, HTs located in regions with higher parallelization, benchmarks such as maxpool_0 and reconfigurable_mvac_0, demonstrate lower detection accuracy, ranging from 55% to 65%. The increased parallelization in these regions leads to a higher density of concurrently active HW blocks, which collectively contribute to elevated baseline power consumption. The resulting noise floor obscures the subtle power fluctuations induced by HT activity, thereby complicating localization. Despite this challenge, the presence of HTs remains detectable, although with reduced accuracy.

For external Trojans such as ext_comb_2, ext_seq_4, and geometric_trojan, which eavesdrop on data between the image buffer and the model, detection accuracy ranges from 36.2% to 72.8%. This variability stems from their interaction with the system: while these pathways are less parallelized than computational layers, detection is still affected by noise levels and the characteristics of the monitored data.

Finally, HT size also influenced accuracy: smaller variants (ht_shft_5_MsCls, ht_shft_1_MsCls) in the same location as ht_shft_10_MsCls were still detectable but with reduced accuracy (~72% vs. 94.4%).

5 Conclusion

The rise of AI accelerators increases security risks, challenging traditional HT detection methods [10]. Detection alone is insufficient for accelerators with highly fluctuating, long power traces. In this work, we proposed a novel approach to localize the regions impacted by HT insertions in AI accelerators using PSCA data. The power traces are segmented based on hardware block activity, treating each region as a word in a sentence representing the full trace. Inspired by the BERT training process, our BERT-style localizer is equipped with an MLM task, enabling it to capture contextual dependencies. This clearly improved test accuracy even when trained with only minimal HTs. This work marks a first step toward models that understand power traces, enabling identification of impacted regions and paving the way for finer-grain, block-level localization. Although scalability and generalization are not explicitly analyzed in this work, they constitute the focus of future investigations. Nevertheless, the proposed method already supports variable-length inputs, enabling characterization of new accelerators.

References

1. Akhoon, M.S., et al.: High performance accelerators for deep neural networks: a review. Expert. Syst. **39**(1), e12831 (2022)
2. Baltagi, A., Nasser, Y., Baghdadi, A.: ML-based hardware trojan detection in AI accelerators via power side-channel analysis. IEEE Trans. Comput. Aided Des. Integr. Circuits Syst. (2025). https://doi.org/10.1109/TCAD.2025.3625109
3. Bhatta, N.P., Amsaad, F.: ML assisted techniques in power side channel analysis for trojan classification. Clust. Comput. **28**(3), 157 (2025)

4. Chen, H., Zhang, X., Huang, K., Koushanfar, F.: AdaTest: reinforcement learning and adaptive sampling for on-chip hardware trojan detection. ACM Trans. Embed. Comput. Syst. **22**(2) (2023). https://doi.org/10.1145/3544015
5. Devlin, J., Chang, M.W., Lee, K., Toutanova, K.: BERT: pre-training of deep bidirectional transformers for language understanding. In: Proceedings of the 2019 conference of the North American chapter of the association for computational linguistics: human language technologies, volume 1 (long and short papers), pp. 4171–4186 (2019)
6. Fan, R., Tang, Y., Sun, H., Liu, J., Li, H.: An efficient ML-based hardware trojan localization framework for RTL security analysis. In: ACM/IEEE International Symposium on Machine Learning for CAD. Association for Computing Machinery, New York, NY, USA (2024). https://doi.org/10.1145/3670474.3685955
7. Guo, C., Yanagisawa, M., Shi, Y.: DSE-based hardware trojan attack for neural network accelerators on FPGAs. IEEE Trans. Neural Netw. Learn. Syst. 1–15 (2024). https://doi.org/10.1109/TNNLS.2024.3482364
8. Hu, W., Chang, C.H., Sengupta, A., Bhunia, S., Kastner, R., Li, H.: An overview of hardware security and trust: threats, countermeasures, and design tools. IEEE Trans. Comput. Aided Des. Integr. Circuits Syst. **40**(6), 1010–1038 (2020)
9. Islam, S.A., Sah, L.K., Katkoori, S.: A framework for hardware trojan vulnerability estimation and localization in RTL designs. J. Hardware Syst. Secur. **4**(3), 246–262 (2020)
10. Jiang, Z., Ding, Q.: A framework for hardware trojan detection based on contrastive learning. Sci. Rep. **14**(1), 30847 (2024)
11. Kang, H., Kang, P.: Transformer-based multivariate time series anomaly detection using inter-variable attention mechanism. Knowl. Based Syst. **290**, 111507 (2024). https://doi.org/10.1016/j.knosys.2024.111507, https://www.sciencedirect.com/science/article/pii/S0950705124001424
12. Kumar, P.: Large language models (LLMs): survey, technical frameworks, and future challenges. Artif. Intell. Rev. **57**(10), 260 (2024)
13. Li, Y., Li, S., Shen, H.: HTrans: transformer-based method for hardware trojan detection and localization. In: IEEE 32nd Asian Test Symposium (ATS), pp. 1–6 (2023). https://doi.org/10.1109/ATS59501.2023.10317971
14. Liu, Z., Hou, J., Wang, J., Yang, C.: A novel two-level protection scheme against hardware trojans on a reconfigurable CNN accelerator. Cryptography **8**(3) (2024). https://doi.org/10.3390/cryptography8030034, https://www.mdpi.com/2410-387X/8/3/34
15. Munir, M., Siddiqui, S.A., Chattha, M.A., Dengel, A., Ahmed, S.: FuseAD: unsupervised anomaly detection in streaming sensors data by fusing statistical and deep learning models. Sensors **19**(11), 2451 (2019)
16. Munir, M., Siddiqui, S.A., Dengel, A., Ahmed, S.: DeepAnT: a deep learning approach for unsupervised anomaly detection in time series. IEEE Access **7**, 1991–2005 (2019). https://doi.org/10.1109/ACCESS.2018.2886457
17. Park, D., Hoshi, Y., Kemp, C.C.: A multimodal anomaly detector for robot-assisted feeding using an LSTM-based variational autoencoder. IEEE Robot. Autom. Lett. **3**(3), 1544–1551 (2018). https://doi.org/10.1109/LRA.2018.2801475
18. Randolph, M., Diehl, W.: Power side-channel attack analysis: a review of 20 years of study for the layman. Cryptography **4**(2) (2020). https://doi.org/10.3390/cryptography4020015, https://www.mdpi.com/2410-387X/4/2/15

19. Sabri, M., Shabani, A., Alizadeh, B.: SAT-based integrated hardware trojan detection and localization approach through path-delay analysis. IEEE Trans. Circuits Syst. II Express Briefs **68**(8), 2850–2854 (2021). https://doi.org/10.1109/TCSII.2021.3074549
20. Su, H., Hu, W., Zhang, X., Zhu, D., Wu, L.: Toward precise and explainable hardware trojan localization at LUT level. IEEE Trans. Comput. Aided Des. Integr. Circuits Syst. **44**(7), 2817–2821 (2025). https://doi.org/10.1109/TCAD.2025.3527377
21. Su, X., Liu, F., Chang, Y., Tanin, E., Sarvi, M., Qi, J.: DualCast: Disentangling Aperiodic Events from Traffic Series with a Dual-Branch Model. arXiv preprint arXiv:2411.18286 (2024)
22. Teledynelecroy: Getting The Most Out of Your Oscilloscope: Setup. https://blog.teledynelecroy.com/2018/01/getting-most-out-of-your-oscilloscope.html (2018)
23. Umuroglu, Y., et al.: FINN: a framework for fast, scalable binarized neural network inference. In: ACM/SIGDA International Symposium on Field-Programmable Gate Arrays, pp. 65–74. FPGA '17, New York, NY, USA (2017). https://doi.org/10.1145/3020078.3021744
24. Vaswani, A., et al.: Attention is all you need. Adv. Neural Inf. Process. Syst. **30** (2017)
25. Wen, Q., et al.: Transformers in time series: A survey. arXiv preprint arXiv:2202.07125 (2022)
26. Yasaei, R., Faezi, S., Al Faruque, M.A.: Golden reference-free hardware trojan localization using graph convolutional network. IEEE Trans. Very Large Scale Integr. (VLSI) Syst. **30**(10), 1401–1411 (2022). https://doi.org/10.1109/TVLSI.2022.3191683
27. Zhang, H., Fan, Z., Zhou, Y., Li, Y.: B-HTRecognizer: bitwise hardware trojan localization using graph attention networks. IEEE Trans. Comput. Aided Des. Integr. Circuits Syst. **44**(6), 2240–2252 (2025). https://doi.org/10.1109/TCAD.2024.3518417
28. Zhang, Y., Yasaei, R., Chen, H., Li, Z., Faruque, M.A.A.: Stealing neural network structure through remote FPGA side-channel analysis. IEEE Trans. Inf. Forensics Secur. **16**, 4377–4388 (2021). https://doi.org/10.1109/TIFS.2021.3106169
29. Zhang, Y., Dong, C., Xu, Y., Yao, Y., Lyu, C.: RLocHT: a hardware trojans localization method utilizing deep learning at the gate-level. In: IEEE 10th Joint International Information Technology and Artificial Intelligence Conference (ITAIC). vol. 10, pp. 2290–2294 (2022). https://doi.org/10.1109/ITAIC54216.2022.9836937

Fault Injection Attacks in On-Chip Interconnects: A Comparative Study of Wishbone, AXI-Lite, and AXI

Hongwei Zhao(✉), Vianney Lapôtre, and Guy Gogniat

Université Bretagne Sud, UMR 6285, Lab-STICC, Lorient, France
{hongwei.zhao,vianney.lapotre,guy.gogniat}@univ-ubs.fr

Abstract. Fault injection attacks exploit physical disturbances to compromise the functionality and security of integrated circuits. As System-on-Chip (SoC) architectures grow in complexity, the vulnerability of on-chip communication buses has become increasingly prominent. Buses, serving as interconnects among various IP cores, represent potential attack vectors for fault-based exploitation. In this study, we perform simulation-driven fault attack injection across three mainstream bus protocols— Wishbone, AXI-Lite, and AXI. We systematically examine fault attack success rates, spatial vulnerability distributions, and timing dependencies to characterize how faults interact with bus-level transactions. The results uncover consistent behavioral patterns across protocols, offering practical insights for both attack modeling and the development of resilient SoC designs.

Keywords: Fault attack injection · System-on-chip · On-chip communication bus · Hardware security · Resilient design

1 Introduction

Embedded systems underpin a wide range of modern applications, from consumer electronics and industrial automation to automotive platforms and critical infrastructure. As their adoption expands into security-critical and real-time domains, ensuring their security and dependability has become increasingly important.

Fault injection attacks have been extensively demonstrated on processors. For example, the Plundervolt attack [10] shows that legitimate voltage scaling interfaces can be exploited to induce faults, successfully compromising Intel SGX enclaves. Complementarily, [3] surveys countermeasures deployed in embedded processors to mitigate such threats and enhance system resilience. Other representative studies, such as [17], demonstrate how fault injection can manipulate the program counter (PC) during data loading, directly subverting instruction flow in ARM-based systems.

Fault injection is also highly relevant in memory subsystems. The work in [18] analyzes attacker capabilities under physical access, discussing memory probing

M. Kowalczyk and C. Monière (Eds.): DASIP 2026, LNCS 16512, pp. 171–183, 2026.
https://doi.org/10.1007/978-3-032-23871-9_14

techniques and corresponding defenses. In [16], the authors show that faulty operations may cause misaddressed memory accesses and propose an access-validation scheme to counter address tampering. More recently, [15] introduces FIPAC, a software-based control-flow integrity (CFI) mechanism operating at basic-block granularity to strengthen execution security against both software and hardware fault injection.

Despite these advances, bus interfaces, which are critical for communication between processing cores, memory, and peripherals, remain comparatively underexplored in the fault resilience literature. This gap is significant, as bus interfaces are shared resources and may be exposed to both internal faults and external physical disturbances.

Recent work such as [8] demonstrates that electromagnetic pulse (EMP) injections targeting bus interfaces can compromise ARM TrustZone security. However, most prior studies focus on individual bus protocols in isolation. In contrast, this work conducts a comparative fault analysis of three widely used on-chip bus protocols—Wishbone, AXI-Lite, and AXI—using simulation-based fault injection campaigns. By stressing these protocols under fault conditions, recurring weaknesses are identified and design guidelines for improving on-chip communication robustness are derived. While [9] studied fault attacks on instruction and data bus interfaces, it neither compared multiple protocols nor systematically explored program-level impacts.

Compared with our research before [19], which focused solely on the Wishbone bus, the present work extends the analysis to AXI-Lite and AXI. The results reveal that all three protocols exhibit structural vulnerabilities exploitable by fault-based attacks. Through systematic simulation-driven campaigns, the study compares targeted register types, data corruption behaviors, and fault success rates across protocols. These observations are abstracted into recurring patterns, leading to a set of design recommendations aimed at enhancing bus-level resilience in embedded systems.

The primary contributions of this paper are summarized as follows:

1. A detailed structural analysis of signal transmission mechanisms across Wishbone, AXI-Lite, and AXI bus protocols;
2. Empirical evaluation of fault injection outcomes across the three buses interfaces using four representative fault models;
3. Insights into how factors such as register types and timing windows affect the success rate of bus-targeted attacks;
4. Practical guidelines for system designers to enhance the robustness of communication interconnects.

The remainder of this paper is structured as follows: Sect. 2 provides a technical overview of the bus protocols and illustrates data flow behaviors before and after fault injection. Section 3 outlines the experimental environment, including the automated fault injection framework and the VerifyPin benchmark. Section 4 presents a comparative vulnerability analysis across the bus protocols under multiple fault models. Finally, Sect. 5 concludes the paper and outlines future research directions.

2 Background

Before performing a fault injection campaign, we examined the Wishbone [11], AXI-Lite [1], and AXI [2] protocols and their associated interfaces in order to identify the control signals that could be manipulated by an attacker. In all three protocols, the transfer process is fundamentally governed by two categories of control signals: access selection (e.g., device or byte selection) issued by the master, and response signals (e.g., completion acknowledgment or error indication) returned by the slave.

In Wishbone, a data transfer is acknowledged by the slave through the `ACK` signal. `ERR` and `RTY` signals optionally indicate an error or a retry request. The `SEL` signal specifies the active byte lanes during a transaction, and the slave selection is determined externally through the address decoding logic.

In AXI-Lite, transfer completion and error reporting are handled by two response channels: `BRESP` for writes and `RRESP` for reads. Each signal provides a two-bit code that indicates whether the access completed successfully (`OKAY`) or resulted in a slave error (`SLVERR`) or decode error (`DECERR`). These responses are synchronized by handshake pairs (`BVALID`/`BREADY` and `RVALID`/`RREADY`), ensuring that both master and slave agree on when a transaction is completed and that the response is valid. During write operations, `WSTRB` defines which byte lanes are active, enabling partial word updates. Slave selection again relies on address decoding within the interconnect rather than a dedicated signal.

In AXI, the same response mechanism is preserved, but additional control signals manage burst transfers. The signals `AWLEN`, `AWSIZE`, and `AWBURST` (and their read equivalents) define the number of data beats, their width, and the burst type. `WLAST` and `RLAST` mark the final data beat of each burst, while `WSTRB` continues to indicate valid bytes on every transfer. Slave selection is still performed through address decoding, but the burst semantics enable efficient block transfers across multiple slaves.

In the implementation considered here, both AXI-Lite and AXI report errors through the `BRESP` and `RRESP` fields. When an error occurs, the slave asserts the corresponding response code and drives `RDATA` to zero, providing an explicit and safe fault indication that prevents invalid data from propagating.

3 Environment Configuration

To perform our experiments, we construct three SoC architectures with the LiteX framework [6]. The target system adopts the RISC-V architecture and integrates a VexRiscv processor core [12]. Three independent SoC configurations are instantiated, each employing a different interconnect protocol—Wishbone, AXI-Lite, and AXI—as the primary communication architecture. The memory contains a ROM for instruction storage, an SRAM for runtime data, a MAIN_RAM region reserved for user-defined memory, and a CSR unit responsible for the UART and timer. All memory blocks are zero-initialized prior to simulation.

To study the effects of fault injection attack in a controlled setting, we employ the VerifyPin benchmark (Sertif project [5]). The benchmark includes eight implementations: an unprotected baseline (V0) and seven versions (V1–V7) with different software countermeasures. We executed all fault-injection campaigns on V0 so that the observed behaviors are not influenced by protection mechanisms. In this program (Listing 1.1), the user input PIN, **g_userPin**, is preset to "0000", whereas **g_cardPin**, is initialized to "4321". The comparison between these two variables simulates a typical failed authentication attempt at line 10. As a result, the flag variable **g_authenticated** is set to "0" at line 12. A fault injection attack is considered successful if it causes **g_authenticated** to be incorrectly set to "1" at line 11, thereby bypassing the authentication logic.

Listing 1.1. C code of VerifyPin function in benchmark V0

```
1  BOOL byteArrayCompare(UBYTE* a1, UBYTE* a2, UBYTE size){
2      for(int i = 0; i < size; i++) {
3          if(a1[i] != a2[i]) return 0;
4      return 1;
5      }
6  }
7  BOOL verifyPIN() {
8      g_authenticated = 0;
9      if(g_ptc > 0) {
10         if(byteArrayCompare(g_userPin, g_cardPin,
               PIN_SIZE) == 1)
11             g_authenticated = 1;
12         else g_authenticated = 0;
13     }
14 }
```

To conduct our fault injection campaign, we utilized Questasim 10.6e as the simulation backend. Given the extensive number of test iterations needed to thoroughly evaluate architectural resilience, we adopted FISSA [13], a Python-based framework designed to automate fault injection campaigns targeting microarchitectural components.

After configuring the simulation environment, we defined attack condition. In line with the benchmark setup, the attacker is assumed to have no knowledge of the correct PIN ("4321") and operates under the assumption that the input PIN ("0000") does not match it. Thus, any successful authentication must arise solely due to fault-induced corruption. For this experiment, we assume an attacker targeting only the control registers connected to the system bus interfaces.

Although this control principle is consistent across protocols, the physical implementation in our LiteX-generated SoC differs from that of a native bus interface. In our design, the AXI-Lite and AXI interfaces are bridged through an internal Wishbone interface via LiteX protocol wrappers. Moreover, the error handling mechanism specifically involves forcibly setting the `ACK` signal to "1" and clearing the output data.

The internal registers of all three bus interfaces are generated by LiteX. We present all targeted registers in Table 1. For Wishbone, fault injection targeted completion, arbitration, and selection logic. For AXI-Lite, representative state registers were selected across functional blocks, such as protocol FSMs, memory and bridge interfaces, selection driver, status flags, interconnect management, and completion flags. For AXI, similar targets were used, with additional burst-related and pipeline control signals absent in AXI-Lite.

Table 1. All the registers targeted by fault injection in the 3 bus interfaces

Bus	Registers
Wishbone	ACK, SEL, done, grant
AXI-Lite	state, selection driver, last_was_read, rr_read_grant, completion flags
AXI	same with AXI-Lite, ax_beat_first, ax_beat_last, last_ar_aw_n, pipe_valid_source

Registers unrelated to bus control—such as those for I/O, LEDs, clock/reset, timers, or holding program data—were excluded from fault injection attacks, as they either bypass control logic or are protected by error-detection mechanisms [7,14], making effective injection unlikely.

To comprehensively evaluate system robustness, we introduced multiple fault models of varying complexity, all operating under a single-cycle fault assumption. In addition to a basic one-bit-flip model, three advanced fault models defined in FISSA were used:

1. **Bit-flip:** One bit-flip in a register.
2. **Manipulate Register:** Bit-flips within a single register. Both the number and location of bit changes are unconstrained, simulating a broad spectrum of register corruption scenarios.
3. **2 Bit-Flips:** Exactly two bit-flips, either within one register or distributed across two.
4. **Manipulate Two Registers:** Injects arbitrary bit-flips into two separate registers simultaneously, capturing more complex situations involving concurrent register corruption.

These attack models can compromise up to four bits of data. This type of attack has been shown to be effective in [4], where fault locations can be precisely controlled using lasers.

Each simulation outcome was categorized into one of four classes:

1. **Crash:** The simulation terminated abnormally due to timeout or triggered crash behavior.
2. **Success:** The authentication logic was bypassed, with `g_authenticated` incorrectly set to "1".
3. **Change:** Memory content was altered, though no authentication success or detection occurred.
4. **Silence:** The fault had no observable effect—no authentication, detection, or state change.

4 Experimental Results

We employed FISSA to automatically generate all TCL commands executed in Quastrasim simulations, targeting registers under our fault models across the three bus interfaces. In total, more than 15 million fault injections were performed. The outcomes were collected in JSON format and post-processed using a Python script. Table 2 summarizes the observed vulnerabilities, reporting the number of occurrences associated with each fault model for Wishbone, AXI-Lite, and AXI.

Table 2. Fault injection results for each bus interface and fault model

Bus	Fault model	Crash	Success	Change	Silence
Wishbone	Bit-flip	47	37	124	2366
	Manipulate Register	320	43	161	6496
	2 Bit-Flips	547	272	914	11137
	Manipulate Two Registers	5178	778	2891	63225
AXI-Lite	Bit-flip	282	4	1913	11577
	Manipulate Register	391	6	2589	29942
	2 Bit-Flips	10725	153	69811	194831
	Manipulate Two Registers	36215	549	224347	1236105
AXI	Bit-flip	158	4	4833	34269
	Manipulate Register	435	6	6723	90996
	2 Bit-Flips	14760	333	428382	1421565
	Manipulate Two Registers	104655	1328	1514455	9762850

The number of simulations increases from Wishbone to AXI-Lite to AXI, reflecting the growing architectural complexity and the larger set of control/status registers. Under simple fault models, Wishbone exhibits a higher rate of successful attacks. This is mainly due to its lightweight handshake mechanism and the absence of advanced protection or reliability features, which makes single faults (e.g., Bit-flip) more likely to directly disrupt transaction control. By contrast, AXI-Lite and AXI incorporate additional handshaking and response channels that tend to absorb or mask such simple faults.

A detailed analysis of the faulted register combinations leading to successful attacks is presented in Tables 3, 4 and 5. For Wishbone (Table 3), most of successful injections target the `ACK` register, regardless of the fault model. The `SEL` register, however, shows particular sensitivity under the Manipulate Register model. For AXI-Lite and AXI (Tables 4 and 5), most successful attacks consistently affect the `state` register, independent of the applied fault model.

Table 3. Distribution of successful register combination attacks under different fault models on the Wishbone bus interface

Fault model	ACK	SEL	ACK & grant	ACK & SEL
Bit-flip	94.59%	5.41%	-	-
Manipulate Register	81.40%	18.60%	-	-
2 Bit-Flips	94.12%	3.68%	1.10%	1.10%
Manipulate Two Registers	85.73%	12.34%	0.38%	1.55%

Table 4. Distribution of successful register combination attacks under different fault models on the AXI-Lite bus interface

Fault model	state	selection driver & state	completion flag & state
Bit-flip	100.00%	-	-
Manipulate Register	100.00%	-	-
2 Bit-Flips	98.02%	1.32%	0.66%
Manipulate Two Registers	98.36%	1.46%	0.18%

Table 5. Distribution of successful register combination attacks under different fault models on the AXI bus interface

Fault model	state	completion flag	completion flag & state
Bit-flip	75.00%	25%	-
Manipulate Register	83.33%	16.67%	-
2 Bit-Flips	60.00%	20%	20.00%
Manipulate Two Registers	83.13%	16.79%	0.08%

4.1 Fault Characterization and Mitigation Approaches

Based on our experimental observations, we categorized the possible fault-induced behaviors for the observed successful attacks across the three bus interfaces into the following classes:

- **Instruction skip:** In a typical SoC, instruction and data fetching rely on handshake synchronization between the CPU and the memory subsystem. The `ACK` signal acknowledges the completion of a read transaction and is directly linked to the CPU's address generation unit to trigger the next fetch cycle. If a transient fault prematurely asserts `ACK`, the CPU may increment the program counter before receiving the correct instruction, causing it to repeatedly execute the previously fetched one. Consequently, the intended instruction—such as the comparison at line 10 of Listing 1.1—is skipped, effectively bypassing the verification logic.
- **Data reset:** The CPU and bus controller maintain internal state registers to manage transaction progress. A fault affecting the `state` register can force

the controller into an unintended transition, activating an error-handling path that drives the data lines to their default value (e.g., zero). As a result, when the CPU samples the data bus, it retrieves an all-zero word for the variable `g_cardPin`, making it identical to `g_userPin` and thereby invalidating the authentication process.
- **Data misread:** Address and byte-lane selection signals determine which memory block and data segment the CPU accesses. A corruption of the `SEL` or address lines can redirect a legitimate read operation toward an unintended memory region. For example, instead of reading `g_userPin` from `SRAM`, the CPU may receive data from the control register space (`CSR`), where `g_cardPin` resides, effectively substituting user data with card data and compromising the verification step.
- **Data multiread:** Under normal operation, decoding logic ensures that only one slave device drives the data bus. A transient fault on the `SEL` or address decoding path may cause multiple slaves (e.g., `CSR` and `SRAM`) to respond simultaneously. The CPU then observes a bus contention and samples a merged or undefined value—the logical combination of multiple sources—corrupting `g_userPin` and potentially matching it with `g_cardPin`.

Based on the previous Tables 3, 4, and 5, our analysis shows that register attacks can succeed through multiple register types.

For Wishbone, the root cause of these vulnerabilities lies in its architectural design. The `ACK` signal directly controls data transfers, yet no timing violation detection is implemented at the bus level. In contrast, the `SEL` signal selects among four memory blocks via a combinational logic, meaning that if compromised, it can arbitrarily redirect data access. Empirical results indicate that `ACK`-related attacks are more prevalent than those targeting `SEL`. The `ACK` signal is register-based and represented by a 4-bit value; flipping any bit can disrupt data transfer, thereby enlarging the attack surface and increasing the probability of success. By comparison, `SEL`-based attacks are more constrained, with their effectiveness heavily dependent on memory initialization states, memory layout strategies, and the behavior defined by the underlying assembly code.

On AXI-lite, the `state`, `selection driver`, and `completion` registers collectively govern the data transfer mechanism, enabling potential attack vectors. The `state` register controls state machine transitions, thereby influencing control signal data and addresses. The `selection driver` affects the selection signal, indirectly determining which memory module the CPU accesses. The `completion` register can influence updates to the `state` register.

Attacks targeting the `state` register exhibit significantly higher success rates than those aimed at the other two registers. This is primarily because the AXI-lite bus interface contains a large number of `state` registers—far exceeding the single `selection driver` and two `completion` registers combined. Moreover, the `state` register can directly force the state machine into error-handling mode, compelling the CPU to read all-zero data. This reset can convert card pins into user pins, thereby contributing to the high success rate.

AXI demonstrates similar characteristics as AXI-lite. The abundance of `state` registers and robust error-handling mechanisms result in consistently high success rates when targeting the `state` registers. `Completion` registers achieve moderate success by indirectly influencing `state` registers. However, due to stricter data control in AXI, compromising the `SEL` signal requires simultaneous manipulation of the `selection driver` and two additional control registers—a condition our fault model cannot meet—resulting in zero successful attacks.

Table 6 presents the proportion of attack effect on instruction and data vulnerabilities observed under the *Manipulate Two Registers* fault model across the three bus interfaces.

Table 6. Percentage of Manipulate Two Registers fault attacks that modify the data or instruction read operation across the three bus interfaces

	Wishbone	AXI-Lite	AXI
data related	17.22%	97.27%	100%
instruction related	82.78%	2.73%	0%

On Wishbone, a high ratio of faults impacts instruction execution. This is primarily due to frequent corruption of instructions such as `lui` and `mv`, which play key roles in data movement and initialization. As these instructions are frequently targeted, the instruction-related fault ratio is notably elevated.

During our experiments, frequent occurrences of data reset were observed on AXI-Lite and AXI, revealing a distinct fault pattern. On Wishbone, when the control signal `done` is enabled, the bus data are initialized to all "1". Since this value does not correspond to any valid variable or instruction, such resets cannot directly satisfy a success condition. By contrast, in AXI-Lite and AXI systems generated by LiteX, two mechanisms may reset the bus data to zero: lightweight error-handling routines and the behavior of state signals. In our benchmark scenario, the user PIN is set to "0000". As a result, a fault-induced reset can lead to interpreting the stored card PIN as "0000" instead of "4321" due to the reset, thereby enabling unauthorized authentication. In these settings, data corruption—rather than instruction corruption—emerges as the dominant attack pathway.

Additionally, across all three bus interfaces, we observed that fault effects for all four injection models consistently occurred during CPU data read phases. This pattern is attributed to the structure of the `VerifyPin` benchmark, which includes very few write operations. Moreover, the registers responsible for handling write transactions on each bus interface are limited in number and infrequently accessed, making them less exposed to injection and thus reducing the likelihood of write-related fault effects.

The proposed classifications serve as a basis for identifying common fault manifestations (i.e., instruction skip, data reset, data misread and data multi-read) and evaluating their severity across different bus interfaces.

Table 7 provides the list of the fault effects for each fault model and for the three buses. Note that the above models can be interpreted differently under certain conditions. For example, if the data reset has no effect, it can be, in certain case, regarded as an instruction skip. Another example is, if the data read operation, due to the fault, leads to "0", then the data multiread model can be treated as a data reset model.

Although the analysis was performed in simulation, mapping these results to silicon requires precise temporal alignment between the injected disturbance and the targeted instruction or bus transaction. Real-world techniques (voltage/clock glitching, EM pulses, focused laser) demand nanosecond-scale timing control and dependable triggers (software markers, side-channels, or debug interfaces); without them, faults seen in simulation are typically probabilistic and poorly reproducible. Hence, the simulated findings identify potential attack vectors, but practical exploitability depends on attacker resources and synchronization capability. Countermeasures should therefore be validated under realistic timing and trigger constraints.

Table 7. Possible fault exploitation for the three different bus interfaces depending on the fault model

	Wishbone	AXI-Lite	AXI
Bit-flip	instruction skip data reset data multiread	data reset	data reset data misread
Manipulate Register	instruction skip data reset data misread data multiread	data reset data misread	data reset data misread
2 Bit-Flips	instruction skip data reset data misread data multiread	instruction skip data reset data misread data multiread	data reset data misread
Manipulate Two Registers	instruction skip data reset data misread data multiread	data reset instruction skip data misread data multiread	data reset data misread

4.2 Design Recommendations for More Resilient Systems

Based on the observed vulnerabilities and fault injection results, we propose the following recommendations for circuit and system designers:

- **Prioritize protection of control registers:** Particular attention should be given to safeguarding data path control elements such as handshake and state registers. Although complex bus protocols inherently resist simplistic fault models due to their layered behavior, they often include a larger number of registers—making them potentially more susceptible to sophisticated or exhaustive fault injection strategies.
- **Reinforce handshake logic:** Faults targeting respond-related signals (e.g., `ACK`) tend to induce instruction skipping with relatively simple fault models, whereas faults on chip-select signals (e.g., `SEL`) typically result in incorrect memory access. Therefore enforced handshake logic warrants stronger protection mechanisms.
- **Go beyond traditional error handling:** Conventional bus-level designs often emphasize error detection and recovery, but may overlook active fault injection scenarios. We recommend that fault tolerance strategies be re-evaluated under adversarial conditions, with particular emphasis on minimizing the propagation effect of fault-handling logic.
- **Mitigate effects of zero-value initialization:** Abnormal inputs such as all-zero patterns should be identified and restricted during critical data flow stages. Sensitive values—such as passwords or authentication codes—should be guarded using masking techniques or write-protection features to prevent forced zeroing. Additionally, authentication processes should incorporate consistency checks to detect and flag suspicious bypass behavior.

In our campaign we evaluated three widely used bus protocols (`Wishbone`, `AXI-Lite`, and `AXI`) using LiteX-derived models to ensure consistent interconnect behavior across experiments. Target selection was systematic rather than ad-hoc: the external fault-injection framework performs parameterized sweeps over register classes (control/status vs. data), address ranges, bit masks, and timing offsets, enabling reproducible coverage across control-register types. We intentionally excluded pure address and data registers because prior work demonstrates effective CPU/memory-level protections (e.g., parity, ECC) for those classes.

5 Conclusion

This work systematically compared Wishbone, AXI-Lite, and AXI under multiple fault models, revealing that while complex buses better resist simple faults, their additional control/status registers create new attack surfaces. Analysis of affected registers, corruption types, and fault timing exposed shared and protocol-specific weaknesses, leading to design recommendations such as protecting handshake signals, hardening against instruction-skips, and refining selection logic. We explicitly note that the experiments were performed in simulation and that mapping simulated faults to FPGA or ASIC implementations requires nanosecond-scale timing alignment and reliable triggers. Real-world techniques impose strict constraints on temporal precision and observability, so many faults trivial to inject in simulation become probabilistic on hardware without privileged triggers (e.g., debug ports or side-channel cues). Future work will therefore focus on extending the fault-injection framework to FPGA platforms and incorporating realistic timing and trigger constraints to assess practical feasibility.

Acknowledgment. This work has been supported by a PhD grant from the DGA Creach Labs research program and the ARED Brittany Region program.

References

1. Arm Limited: AMBA AXI and ACE protocol specification, ARM IHI 0022E. Arm (2013). https://documentation-service.arm.com/static/5f915b62f86e16515cdc3b1c
2. Arm Limited: AMBA AXI4-Lite Interface (subset of AXI4). Arm (2013). https://documentation-service.arm.com/static/5f915b62f86e16515cdc3b1c. In: AMBA AXI and ACE Protocol Specification, ARM IHI 0022E
3. Boulifa, R., Di Natale, G., Maistri, P.: Countermeasures against fault injection attacks in processors: a review. Information (2078-2489) **16**(4) (2025)
4. Colombier, B., et al.: Multi-spot laser fault injection setup: new possibilities for fault injection attacks. In: International Conference on Smart Card Research and Advanced Applications, pp. 151–166. Springer (2021). https://doi.org/10.1007/978-3-030-97348-3_9
5. Dureuil, L., Petiot, G., Potet, M.L., Le, T.H., Crohen, A., de Choudens, P.: FISSC: a fault injection and simulation secure collection. In: Computer Safety, Reliability, and Security: 35th International Conference, SAFECOMP 2016, Trondheim, Norway, 21-23 September 2016, Proceedings 35. pp. 3–11. Springer (2016). https://doi.org/10.1007/978-3-319-45477-1_1
6. Kermarrec, F., Bourdeauducq, S., Lann, J.C.L., Badier, H.: LITEX: an open-source soc builder and library based on MIGEN python DSL. arXiv preprint arXiv:2005.02506 (2020)
7. Kiaei, P., Mercadier, D., Dagand, P.E., Heydemann, K., Schaumont, P.: Custom instruction support for modular defense against side-channel and fault attacks. In: International Workshop on Constructive Side-Channel Analysis and Secure Design, pp. 221–253. Springer (2020). https://doi.org/10.1007/978-3-030-68773-1_11
8. Mishra, N., Chakraborty, A., Mukhopadhyay, D.: Faults in our bus: Novel bus fault attack to break arm trustzone. In: Network and Distributed System Security Symposium (2024)
9. Moro, N., Dehbaoui, A., Heydemann, K., Robisson, B., Encrenaz, E.: Electromagnetic fault injection: towards a fault model on a 32-bit microcontroller. In: 2013 Workshop on Fault Diagnosis and Tolerance in Cryptography, pp. 77–88 (2013). https://doi.org/10.1109/FDTC.2013.9
10. Murdock, K., Oswald, D., Garcia, F.D., Van Bulck, J., Gruss, D., Piessens, F.: PlunderVolt: software-based fault injection attacks against intel SGX. In: 2020 IEEE Symposium on Security and Privacy (SP), pp. 1466–1482 (2020). https://doi.org/10.1109/SP40000.2020.00057
11. OpenCores Organization: WISHBONE System-on-Chip (SoC) interconnection architecture for portable IP cores, Rev. B.3. OpenCores (2002). https://cdn.opencores.org/downloads/wbspec_b3.pdf
12. Papon, C.: VEXRISCV: a modular RISC-V CPU generator (2018). https://github.com/SpinalHDL/VexRiscv. Accessed 1 Apr 2025

13. Pensec, W., Lapôtre, V., Gogniat, G.: Scripting the unpredictable: automate fault injection in RTL simulation for vulnerability assessment. In: 2024 27th Euromicro Conference on Digital System Design (DSD), pp. 369–376. IEEE (2024)
14. Sakamoto, J., Hayashi, S., Fujimoto, D., Matsumoto, T.: How to code data integrity verification secure against single-spot-laser-induced instruction manipulation attacks. In: 2020 IEEE/ACS 17th International Conference on Computer Systems and Applications (AICCSA), pp. 1–8 (2020). https://doi.org/10.1109/AICCSA50499.2020.9316521
15. Schilling, R., Nasahl, P., Mangard, S.: FIPAC: Thwarting fault-and software-induced control-flow attacks with arm pointer authentication. In: International Workshop on Constructive Side-Channel Analysis and Secure Design, pp. 100–124. Springer (2022). https://doi.org/10.1007/978-3-030-99766-3_5
16. Schilling, R., Werner, M., Nasahl, P., Mangard, S.: Pointing in the right direction - securing memory accesses in a faulty world. In: Proceedings of the 34th Annual Computer Security Applications Conference, pp. 595–604 (2018). https://doi.org/10.1145/3274694.3274728
17. Timmers, N., Spruyt, A., Witteman, M.: Controlling PC on arm using fault injection. In: 2016 Workshop on Fault Diagnosis and Tolerance in Cryptography (FDTC), pp. 25–35 (2016). https://doi.org/10.1109/FDTC.2016.18
18. Werner, M., Schilling, R., Unterluggauer, T., Mangard, S.: Protecting RISC-V processors against physical attacks. In: 2019 Design, Automation and Test in Europe Conference and Exhibition (DATE), pp. 1136–1141 (2019). https://doi.org/10.23919/DATE.2019.8714811
19. Zhao, H., Lapotre, V., Gogniat, G.: Communication architecture under siege: an in-depth analysis of fault attack vulnerabilities and countermeasures. In: 2024 IEEE International Conference on Cyber Security and Resilience (CSR), pp. 890–896. IEEE (2024)

UAVLnQ: An Architecture for Security Analysis of Cyber-Physical Network Behavior in UAV Swarms

Abdelrahman Yousef[1(✉)], Chinya Tsai[1], Nikita Nilesh Mistry[2], Maria Méndez Real[3], and Guy Gogniat[3]

[1] Université Bretagne Sud, Lorient, France
yousef.e2405882@etud.univ-ubs.fr
[2] Department of Computer Science and Engineering, College of Engineering, Qatar University, Doha, Qatar
[3] UMR 6285, Lab-STICC, Université Bretagne-Sud, Lorient, France

Abstract. The design and validation of security architectures for Unmanned Aerial Vehicle (UAV) swarms, such as on-board Intrusion Detection Systems (IDS), require simulation platforms that can rapidly prototype complex cyber-physical scenarios. These platforms must simultaneously capture accurate flight dynamics and realistic network behavior while remaining scalable and reproducible. This paper introduces Unmanned Aerial Vehicle Link and Queue (UAVLnQ), an open-source architecture explicitly designed to address these challenges. It integrates ArduPilot Software-in-the-Loop (SITL) with ns-3 through a ZMQ middleware that guarantees synchronization between mobility control and packet-level events. UAVLnQ features an autopilot mission control system for defining and executing complex swarm behaviors. We demonstrate its capabilities in a leader-follower formation under representative cyber-attacks—including command injection, spoofing, and denial of service—providing a validated environment for the development and testing of novel security and resilience solutions. UAVLnQ thus establishes itself as a reproducible and extensible foundation for a new generation of UAV swarm networking and cyber-defense research.

Keywords: Unmanned Aerial Vehicles · ns-3 · ArduPilot · MAVLink · UAV Swarm · UAV Attacks

1 Introduction

Unmanned Aerial Vehicle (UAV) swarms are increasingly characterized by autonomous and cooperative operations in which mission success depends on reliable and resilient wireless mesh networking [3,23]. Such swarms are now deployed in safety-critical applications including emergency response, surveillance, and infrastructure inspection, where communication failures or malicious

M. Kowalczyk and C. Monière (Eds.): DASIP 2026, LNCS 16512, pp. 184–196, 2026.
https://doi.org/10.1007/978-3-032-23871-9_15

interference can rapidly propagate into unsafe physical behavior or mission failure. However, this same reliance on wireless communication introduces a substantial attack surface; adversarial actors may exploit vulnerabilities through command injection, GPS spoofing, or denial-of-service (DoS) attacks, leading to loss of control, trajectory deviation, or forced termination of swarm missions [3, 10].

As swarm coordination logic and autonomy grow in complexity, there is an increasing need for simulation platforms that accurately model the cyber–physical coupling between network-layer events and UAVs [9, 20]. In particular, the validation of Intrusion Detection Systems (IDS) depends on preserving the causal relationship between network anomalies and physical responses, as decoupled simulations can lead to misleading assessments of system resilience [20].

Direct experimentation with cyber-physical UAV attacks is often infeasible due to safety concerns, regulatory constraints, and legal restrictions on jamming and spoofing in open airspace [6, 20]. Consequently, simulation environments have become the primary means for developing and validating UAV security mechanisms. However, many existing simulation frameworks fail to preserve the fidelity required for cyber-physical security analysis [9]. Robotics-centric simulators such as Gazebo abstract communication into idealized inter-process messaging, preventing the evaluation of protocol-level exploits [13]. Hybrid co-simulation frameworks, most notably FlyNetSim [5, 13], integrate ArduPilot with ns-3 but often bypass the simulated network for mobility or position updates to improve scalability. This decoupling can produce a false sense of resilience: a simulated denial-of-service attack may saturate the network, yet UAVs continue executing their missions because control inputs are delivered out-of-band.

Similarly, lightweight frameworks such as DroNS-3 emphasize ease of use and high-level mission logic but rely on abstractions that obscure low-level packet dynamics and timing behavior. While such tools are valuable for algorithm prototyping, they are ill-suited for validating IDS mechanisms where packet-level visibility and causal cyber-physical effects are essential [6, 11]. As a result, existing platforms struggle to support reproducible security experiments in which network-layer attacks induce realistic and observable physical consequences. This work introduces UAVLnQ, an open-source cyber-physical co-simulation architecture explicitly designed to close this validity gap. Unlike existing frameworks, UAVLnQ enforces a strict coupling between communication and mobility by ensuring that all MAVLink control, telemetry, and inter-UAV coordination traffic traverses the simulated wireless network. Any network degradation or malicious manipulation therefore directly influences UAV flight behavior.

The primary objectives of this work are fourfold:

- To accurately model UAV flight dynamics and mission execution using ArduPilot Software-in-the-Loop (SITL) while preserving native MAVLink communication semantics;
- To simulate realistic wireless networking conditions in which latency, packet loss, and adversarial traffic causally propagate into UAV behavior;

- To support configurable multi-UAV swarm topologies with inter-UAV communication and coordination; and
- To provide a reproducible adversarial experimentation framework for UAV cybersecurity and Intrusion Detection System research.

UAVLnQ achieves these objectives by tightly integrating ArduPilot SITL with ns-3 through a lightweight ZeroMQ middleware that preserves the MAVLink communication pipeline while enabling low-latency, bidirectional message exchange. The framework produces synchronized packet-level network traces alongside timestamped ArduPilot telemetry logs, enabling the generation of high-fidelity cyber-physical datasets suitable for IDS development and security evaluation. UAVLnQ is fully open source and publicly available [25].

We validate the proposed architecture using a leader–follower swarm topology in which the leader UAV acts as a swarm controller, issuing MAVLink commands to follower UAVs over the simulated network. Scalability is evaluated up to 20 UAVs with stable resource usage and end-to-end latency below 100 ms. Representative cyber-attack scenarios including command injection, spoofing, and DoS are introduced to demonstrate how network-layer exploits propagate into observable flight anomalies. These results establish UAVLnQ as a reliable and extensible foundation for UAV swarm networking and cyber-defense research.

2 Related Work

UAV simulation environments are broadly categorized into robotics-centric, network-centric, and integrated co-simulation architectures.

2.1 Robotics and Digital Twin Simulators

General-purpose robotics simulators are frequently used to model UAVs because they provide the necessary physics engines to replicate flight dynamics and aerodynamic forces. Platforms such as Gazebo [17] and AirSim [21] excel at high-fidelity physical modeling and photorealistic rendering. However, they abstract communication into idealized inter-process messaging, neglecting stochastic network phenomena—such as packet loss, jitter, and protocol overhead—which are foundational for effective security research. Recent Digital Twin implementations, notably GazeboNS3 [8], extend these platforms by integrating battery models and computational constraints. GazeboNS3 is highly effective for optimizing the energy profile of specific hardware, yet its dependency on full physics rendering limits scalability. Benchmarks indicate that real-time performance in such systems drops noticeably as node counts rise above ten. Additionally, these systems often lack native modules for adversarial packet injection, requiring extensive modification to support cybersecurity workflows. Trace-based tools like DroneNet-Sim [18] offer an alternative by replaying recorded mobility traces. While computationally efficient, trace-based systems cannot react dynamically to attacks—a drone will not stop when a command is jammed if it is following a pre-recorded path—rendering them unsuitable for active defense validation.

2.2 Network-Centric Simulators

Discrete-event network simulators, such as ns-3, OMNeT++ [22], and OPNET, excel at modeling the lower layers of the protocol stack, offering granular control over channel interference, propagation loss, and routing logic [5]. However, these platforms treat mobile nodes as abstract entities governed by simplified kinematic models (e.g., Random Waypoint or Gauss-Markov) rather than dynamic physical bodies [5]. They lack the aerodynamic engines necessary to simulate inertia, wind resistance, or thrust-to-weight ratios. Consequently, standalone network simulators cannot capture the cyber-physical feedback loop essential for security validation; they cannot model scenarios where network latency induces mechanical instability or where a kinetic maneuver inadvertently disrupts the antenna link budget. While recent Python-based extensions like UavNetSim-v1 [27] attempt to introduce mobility features, they lack the real-time synchronization with flight controllers required to test active autopilot defense mechanisms.

2.3 Integrated Co-simulation Architectures

Co-simulation frameworks link distinct simulators to address networking fidelity, though often with significant trade-offs. FlyNetSim [5] pioneered the ArduPilot/ns-3 integration via middleware. However, it's reported scalability is measured without accounting for full intra-swarm mesh communication, making its statistics unsuitable for rigorous cybersecurity validation where coordinated, network-dependent attacks must propagate through the simulated channel.

While frameworks such as ROS-NetSim [7], FANET-Sim [24] and Cornet [2] provide robust co-simulation capabilities, their dependence on the Robot Operating System (ROS) [19] introduces substantial computational overhead that constrains scalability for large swarms. In contrast, DroNS-3 offers a streamlined interface between ArduPilot and ns-3 tailored for educational use and single-agent logic. However, its reliance on the DroneKit [1] API abstracts critical low-level packet dynamics, obscuring the data granularity necessary for validating Intrusion Detection Systems (IDS). Other approaches, such as CUSCUS [26], leverage Linux containers to bridge simulators but remain proprietary and closed-source. UAVLnQ improves upon these state-of-the-art methods by implementing a ZMQ middleware, which guarantees tight cyber-physical coupling without the resource penalties of ROS or heavy physics engines, thereby optimizing the system for scalable swarm simulation and adversarial injection.

3 UAVLnQ Framework

The UAVLnQ framework integrates several core components into a single simulation environment. At its foundation lies a flight simulator that provides realistic Unmanned Aerial Vehicle (UAV) dynamics and mission control logic, coupled with a network simulator that models communication links at the packet level. A middleware layer connects these two domains, forming a closed-loop architecture for the analysis.

3.1 Core Components

To achieve realistic flight dynamics and mission control, UAVLnQ integrates ArduPilot SITL [4] UAVs, which are visualized in 3D via Gazebo [17] or in 2D via QGroundControl [15], and the DroneKit API [1] mission planning and controller for high-level mission logic, while ns-3 [16], a discrete-event network simulator, models packet level communication across the entire swarm topology including inter-UAV links and ground control station connectivity.

Since these components operate as separate processes, we connect them using a ZMQ [12] middleware that supports bidirectional message exchange. This publish-subscribe architecture creates end-to-end data paths for mission commands, telemetry data, and network packets.

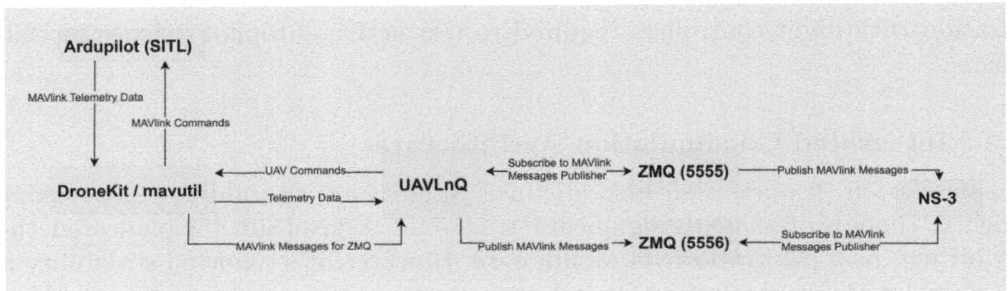

Fig. 1. UAVLnQ Framework.

3.2 Overall Architecture

Figure 1 illustrates the overall architecture of the UAVLnQ framework. The mission control logic, implemented in Python, serves as the central coordinator, managing connections and mission execution for all UAV instances while logging telemetry data. Communication is handled via two primary ZMQ sockets (ports 5555 and 5556) that bridge MAVLink traffic between ArduPilot/mavutil components and the ns-3 simulation which ensures that MAVLink messages from ArduPilot reach ns-3 for network propagation, while packets generated in ns-3 are delivered back to the corresponding ArduPilot instances for execution. Unlike approaches based on high-overhead visualizers such as AirSim, UAVLnQ employs a ZMQ bridge that acts as a low-latency pass-through between simulators. Through direct one-to-one mapping between UAV instances and ns-3 nodes, its design leverages ns-3's inherent capability to handle complex topologies while capturing comprehensive network traffic traces of the entire swarm topology. The multi-UAV simulation is configured via a single JSON file that defines per-UAV parameters including DroneKit connection endpoints, MAVLink communication and parser ports, QGroundControl visualization ports and the path of the ns-3 network simulation script, allowing straightforward swarm scaling by adding configuration entries. This architecture enables UAVLnQ to be customizable and scale to large swarm systems.

Table 1. UAVLnQ Commands

Command	Description
MoveToWaypoint	Move the UAV to specified local coordinates.
Sleep	Pause execution of commands for a specified duration.
ReturnHome	Return to the launch location.
Land	Land at the current UAV position.

4 UAVLnQ Autopilot

The autopilot architecture shown in Fig. 2 comprises five modular components working in concert to manage swarm operations. The DroneCommander class coordinates all UAV connections by reading the JSON configuration file to obtain per-UAV connection parameters, establishing and managing parallel connections via DroneKit while verifying armable state before mission initiation. It publishes MAVLink messages to ns-3 through ZMQ, launches the network simulator, and monitors mission files for dynamic command updates. Individual DynamicMissionController instances manage per-UAV command execution, where commands are defined through a base class interface that issues instructions to ArduPilot SITL via the DroneKit API. Table 1 provides an overview of available UAV commands. The DataLogger class captures telemetry data and stores it in timestamped CSV files, while the MAVLink Parser bridges network traffic between ns-3 and autopilot instances. This modular architecture allows rapid prototyping of swarm missions and cyber attack scenarios with low implementation complexity and provides the flexibility to dynamically modify mission behavior during execution.

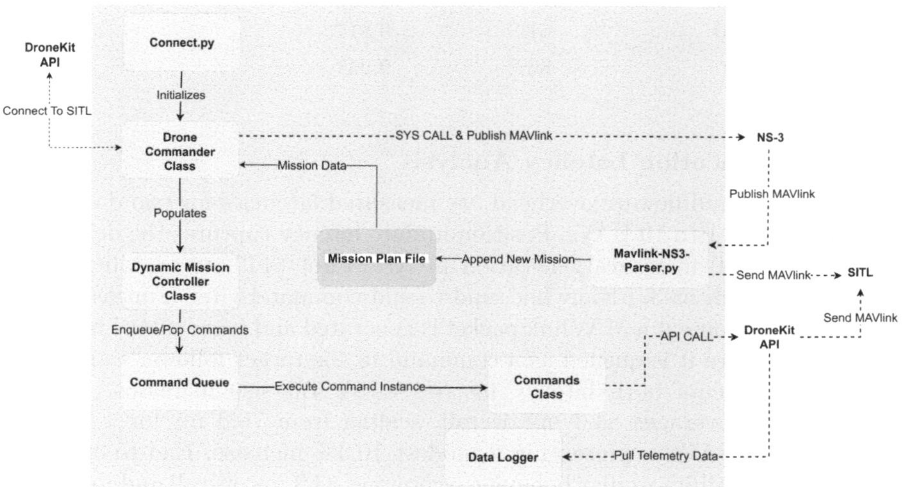

Fig. 2. UAVLnQ Autopilot Structure.

5 Evaluation Analysis

To assess UAVLnQ's scalability, computational efficiency, and communication latency characteristics, we evaluate the framework across different swarm sizes by executing standardized mission scenarios. All experiments were conducted on an AMD Ryzen 5 7520U (4 cores, 8 threads, 4.4 GHz max) with 16 GB RAM running Ubuntu 24.04 LTS. The evaluation quantifies computational overhead as swarm size increases and characterizes the ZMQ-based middleware latency for both position updates and command propagation between flight simulation and network modeling.

5.1 Computational Overhead and Scalability

CPU utilization and memory consumption was evaluated across swarm sizes of 3, 10, and 20 UAVs executing 300-s missions. Each configuration was tested through 10 independent runs to ensure statistical reliability, with results averaged across repetitions. Table 2 shows that CPU utilization scales approximately linearly with swarm size, increasing from 29.8% (3 UAVs) to 61.8% (10 UAVs) and 86.2% (20 UAVs), indicating CPU becomes the primary bottleneck. Memory consumption remains stable at approximately 9.5–10 GB across all configurations, dominated by fixed overhead from ns-3 and QGroundControl rather than per-UAV costs. The framework maintains efficient execution up to 10 UAVs with comfortable headroom (62% CPU), while 20 UAVs approaches saturation at 86% CPU.

Table 2. UAVLnQ Computational Overhead

Number of UAVs	Avg. CPU (%)	Avg. Memory (MB)
3	29.8	9,565
10	61.8	9,817
20	86.2	9,947

5.2 Communication Latency Analysis

To characterize middleware overhead, we measured latencies for two data paths across swarms of 1 to 10 UAVs. Position update latency captures the delay from GPS_RAW_INT messages generation in ArduPilot SITL to mobility model synchronization in ns-3. Meanwhile, end-to-end command latency measures the delay from the moment a MAVlink packet is generated and transmitted from ns-3 to the point where it is queued as a command in the target follower's autopilot.

Figure 3 presents both latency metrics as swarm size increases. Position update latency averages 84.3 ms overall, scaling from 76.5 ms for 1 UAV to 84.4 ms for 10 UAVs, representing a modest 10.4% increase. End-to-end command latency exhibits similar behavior, averaging 83.3 ms overall and increasing from 79.8 ms to 83.3 ms across the same range, a 4.4% increase. Both metrics remain consistently below 100 ms across all tested swarm sizes.

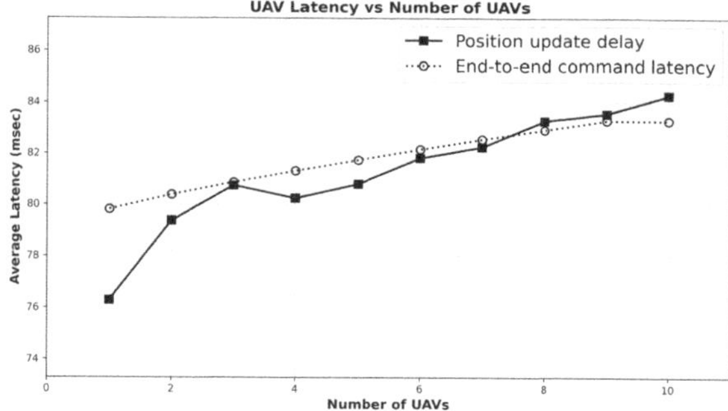

Fig. 3. Communication latency vs. swarm size. Position update delay and end-to-end command latency.

6 Use Cases

We demonstrate UAVLnQ's capabilities through two representative scenarios: (i) a baseline mission operation under nominal conditions, and (ii) the same topology subjected to cyberattacks. These scenarios validate the framework's ability to accurately reproduce UAV mission execution while systematically modeling adversarial interference within a controlled and reproducible environment, generating correlated datasets.

6.1 Case I: Normal Mission Operation

The first use case comprises three UAV nodes in a fully connected mesh network shown in Fig. 4a. Node mobility is synchronized with ArduPilot Software-inthe-Loop (SITL) via `GPS_RAW_INT` messages over ZMQ, while communication follows the MAVLink protocol [14]: the leader sends `MISSION_ITEM` commands to followers, and all UAVs broadcast telemetry for swarm-wide awareness.

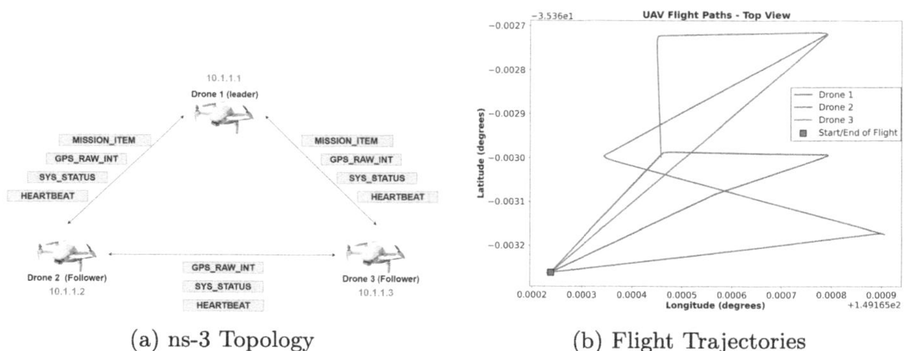

(a) ns-3 Topology

(b) Flight Trajectories

Fig. 4. Baseline mission: network topology and flight paths.

Figure 4b shows the mission flight path of each UAV as the leader executes its mission and dispatches waypoints to followers at scheduled intervals. Upon leader landing, all followers execute Return-to-Launch for coordinated mission termination.

Generated Dataset. Table 3 summarizes the dataset generated from the baseline mission. Each UAV produces telemetry logs sampled at 0.5-second intervals, while ns-3 captures complete packet level network traces.

Table 3. Case I Dataset Statistics

Metric	Value
Mission Duration	200 s
Telemetry CSV Samples per UAV	400 samples
Total Network Packets	25,000 packets

6.2 Case II: Normal Mission Operation Under Cyber Attacks

Extending the first use case. A malicious node was introduced in the same network topology to intercept unencrypted MAVLink traffic and inject crafted packets to both UAVs and ground station to evaluate the adversarial behaviors and their effects on flight dynamics.

6.3 Attack Categories

UAVLnQ supports prototyping of several classes of UAV attacks using different MAVlink messages. We validated four representative categories frequently considered in UAV security research:

- **Command Injection:** Injected spoofed commands including `MISSION_ITEM` to modify mission waypoints, `MAV_CMD_COMPONENT_ARM_DISARM` for forced disarm, `MAV_CMD_DO_FLIGHTTERMINATION` for full flight termination, and `MAV_CMD_DO_CHANGE_SPEED` to change flight speed.
- **Spoofing:** Falsified `GPS_RAW_INT` packets distorting localization and manipulated `BATTERY_STATUS` triggering unintended failsafes.
- **Navigation Hijacking:** Spoofed `MAV_CMD_DO_SET_HOME` commands to alter home position, thereby redirecting the Return-to-Launch (RTL) behavior.
- **Denial of Service (DoS) Attacks:** Flood the communication channels with malicious traffic, including `HEARTBEAT` flooding and ghost drone flooding (simultaneous injection of `GLOBAL_POSITION_INT`, `HEARTBEAT`, `SYS_STATUS`, and `GPS_RAW_INT` to create fictitious UAVs).

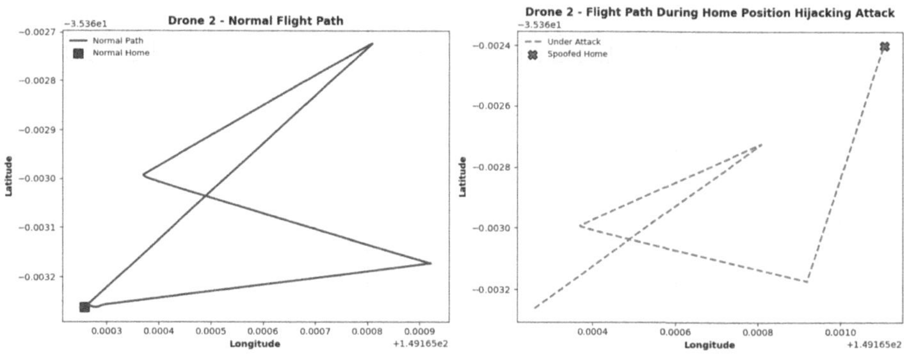

Fig. 5. 2nd UAV trajectory comparison: Normal mission vs Home Position Hijacking Attack.

6.4 Scenario I: Navigation Hijacking

The attacker starts injecting spoofed `MAV_CMD_DO_SET_HOME` commands into the simulation, targeting 2nd UAV and repeatedly redirecting its home position to a point 124 m far from the original launch site. The attack function reschedules itself every second, creating a persistent home position spoofing flood. Figure 5 illustrates the resulting trajectory divergence: under normal conditions, the UAV completes its waypoints and returns to the original home location; under attack, it diverts to the hijacked coordinates upon receiving the return-to-launch command. The redirected path increases the total flight distance from 247 m to 275 m, while the mission duration decreases to 177 s because the spoofed home location is closer to the final waypoint. Table 4 summarizes the corresponding dataset characteristics.

Table 4. Dataset Comparison for 2nd UAV: Normal vs Home Position Hijacking Attack

Metric	Normal	Attack
Mission Duration	201 s	177 s
Path Distance	247 m	275 m
End Position Deviation	0 m	124 m
Avg telemetry samples per UAV	400 samples	353 samples
Total Network Packets	25,000 packets	27,332 packets

6.5 Scenario II: Operational Efficiency Degradation

The third scenario demonstrates operational degradation through speed manipulation. The attacker injects `MAV_CMD_DO_CHANGE_SPEED` commands every 5 s, reducing second UAV ground speed to 0.5 m/s. Table 5 shows mission duration extended from 201 s to 303 s, while average ground speed decreased from 1.25 m/s to 0.80 m/s. Table 5 also compares dataset characteristics.

Table 5. Dataset Comparison for the 2nd UAV: Normal vs Speed Manipulation Attack

Metric	Normal	Attack
Mission Duration	201 s	303 s
Avg Ground Speed	1.25 m/s	0.80 m/s
Avg telemetry samples per UAV	400 samples	605 samples
Total Network Packets	25,000 packets	44,477 packets

7 Conclusion

UAV swarms are becoming essential in emergency response, logistics, and security, yet their development is limited by the lack of simulation platforms that simultaneously capture realistic flight dynamics, packet-level network behavior, and cyber-physical attack propagation. UAVLnQ addresses this gap by integrating ArduPilot SITL with ns-3 through a ZMQ middleware, enabling synchronized mobility and communication modeling. As a general-purpose and extensible simulator, UAVLnQ supports rapid prototyping of swarm missions and attack scenarios with low implementation complexity. The framework's modular architecture, driven by JSON configuration and command-based mission control, enables straightforward adaptation for research and education and extends beyond multirotors to fixed-wing aircraft and ground vehicles. Our evaluation demonstrated UAVLnQ's computational efficiency and scalability, with stable memory consumption and communication latency remaining consistently below 100 ms for both position updates and command propagation. We validated the framework's cyber-physical testbed capabilities through two representative attack scenarios—navigation hijacking and operational efficiency degradation to demonstrate observable consequences in both network traces and flight telemetry. Each scenario generates correlated datasets combining pcap network captures and CSV telemetry logs, providing data that can be used for instance to build and test IDS models. Compared to existing solutions, UAVLnQ provides an efficient platform for simulating realistic cyber-physical behavior across large UAV swarms. The complete source code and tool is released as open source [25] and our hope is that it will serve as a basis for the development of new contributions in this direction. Future work will expand the framework with additional attack models, defensive mechanisms, and more complex swarm topology and coordination strategies, alongside developing lightweight IDS solutions deployable on-board UAVs.

Acknowledgments. This work was carried out as part of the CYBERUS Erasmus Mundus Master's programme. The authors would like to thank the programme for its support.

References

1. 3DRobotics Inc.: DroneKit developer tools for drones (2023). https://dronekit.io/
2. Acharya, S., Bharadwaj, A., Simmhan, Y., Gopalan, A., Parag, P., Tyagi, H.: CORNET: a co-simulation middleware for robot networks. In: 2020 International Conference on COMmunication Systems & NETworkS (COMSNETS), pp. 245–251 (2020)
3. Andreoni, M., Baddeley, M., Lunardi, W.T., Pandey, A., Giacalone, J.P.: Towards secure wireless mesh networks for UAV swarm connectivity: current threats, research, and opportunities (2021)
4. ArduPilot Development Team: ArduPilot SITL (software in the loop). https://ardupilot.org/dev/docs/sitl-simulator-software-in-the-loop.html. Accessed: 09 Sep 2025
5. Baidya, S., Shaikh, Z., Levorato, M.: FlyNetSim: an open source synchronized UAV network simulator based on ns-3 and ArduPilot. In: Proceedings of the 21st ACM International Conference on Modeling, Analysis and Simulation of Wireless and Mobile Systems, pp. 37–45. ACM (2018)
6. Burhan Mohammed, A., Fourati, L., Fakhrudeen, A.: Comprehensive systematic review of intelligent approaches in UAV-based intrusion detection, blockchain, and network security. Comput. Netw. **239**, 110140 (12 2023)
7. Calvo-Fullana, M., Mox, D., Pyattaev, A., Fink, J., Kumar, V., Ribeiro, A.: ROS-NetSim: a framework for the integration of robotic and network simulators. IEEE Robot. Autom. Lett., 1 (2021)
8. Cauwe, D., Wang, C., Han, Q.: GazeboNS3: a digital twin system for UAV swarms, pp. 699–704. Association for Computing Machinery, New York, NY, USA (2025)
9. Meheretu, S.E., Nigussie, E., Gebremeskel, G.B., Hailesilassie, S.Y.: A systematic literature review on spoofing and jamming approaches in unmanned aerial vehicles navigation. J. Aerosp. Technol. Manag. **17**, e3425 (2025)
10. Ceviz, O., Sadioğlu, P., Sen, S.: A survey of security in UAVs and FANETs: Issues, threats, analysis of attacks, and solutions. arXiv preprint arXiv:2306.14281 (06 2023)
11. Hassler, S.C., Mughal, U.A., Ismail, M.: Cyber-physical intrusion detection system for unmanned aerial vehicles. IEEE Trans. Intell. Transp. Syst. **25**(6), 6106–6117 (2024)
12. iMatix Corporation: ZeroMQ (ZMQ) high-performance messaging library. https://zeromq.org/. Accessed 09 Sep 2025
13. Jiang, W., Han, H., He, M., Gu, W.: Network simulation tools for unmanned aerial vehicle communications: a survey. Int. J. Commun Syst **37**(15), e5878 (2024)
14. MAVLink development team: MAVLink micro air vehicle communication protocol. https://mavlink.io/en/. Accessed 09 Sep 2025
15. Meier, L., QGroundControl development team: QGroundControl: open source ground control station for UAVs. https://qgroundcontrol.com/. Accessed 09 Sep 2025
16. ns-3 Consortium: ns-3 network simulator. https://www.nsnam.org/. Accessed 09 Sep 2025
17. Open robotics: Gazebo robotics simulator. https://gazebosim.org/. Accessed 09 Sep 2025
18. Qu, C., Morel, A.E., Dahlquist, D., Calyam, P.: DroneNet-Sim: a learning-based trace simulation framework for control networking in drone video analytics. In: Proceedings of the 6th ACM Workshop on Micro Aerial Vehicle Networks, Systems,

and Applications. DroNet 2020, Association for Computing Machinery, New York, NY, USA (2020)

19. Quigley, M., et al.: ROS: an open-source robot operating system. In: ICRA Workshop on Open Source Software, vol. 3 (2009)
20. Rahut, R.: Countering unauthorized drone activities: an analysis of drone jamming techniques, challenges, and ethical implications. Master's thesis, LAB University of Applied Sciences, Lappeenranta, Finland (2024). https://urn.fi/URN:NBN:fi:amk-2024121335580
21. Shah, S., Dey, D., Lovett, C., Kapoor, A.: AirSim: high-fidelity visual and physical simulation for autonomous vehicles. In: Field and Service Robotics: Results of the 11th International Conference, pp. 621–635. Springer (2017). https://doi.org/10.1007/978-3-319-67361-5_40
22. Varga, A., Hornig, R.: OMNeT++ discrete event simulator. https://omnetpp.org/. Accessed 09 Sep 2025
23. Wang, X., Zhao, Z., Yi, L., Guo, L., Yu, F.R., Guo, S.: A survey on security of UAV swarm networks: attacks and countermeasures. ACM Comput. Surv. **57** (2024)
24. Xie, C., Li, Y., Zhang, T., Zhang, C.: FANET-sim: a simulation framework for UAV swarm communication. In: Proceedings of the 2025 5th International Conference on Computer Network Security and Software Engineering, pp. 183–189. CNSSE 2025, Association for Computing Machinery, New York, NY, USA (2025)
25. Yousef, A., Tsai, C., Mistry, N.N., Méndez Real, M., Gogniat, G.: UAVLnQ: an architecture for security analysis of cyber-physical network behavior in UAV swarms. https://github.com/Wh02m1/UAVLnQ
26. Zema, N.R., Trotta, A., Natalizio, E., Di Felice, M., Bononi, L.: The CUSCUS simulator for distributed networked control systems: architecture and use-cases. Ad Hoc Netw. **68**, 33–47 (2018)
27. Zhou, Z., et al.: UavNetSim-v1: a Python-based simulation platform for UAV communication networks (2025)

Author Index

M. Kowalczyk and C. Monière (Eds.): DASIP 2026, LNCS 16512, p. 197, 2026.
https://doi.org/10.1007/978-3-032-23871-9

The manufacturer's authorised representative in the EU is Springer Nature Customer Service Centre GmbH, Europaplatz 3, 69115 Heidelberg, Germany. If you have any concerns regarding our products, please contact ProductSafety@springernature.com

Printed and bound by CPI Group (UK) Ltd, Croydon, CR0 4YY
07/07/2026
02160913-0004